World Statistics on Mining and Utilities

2018

Statistics Division
United Nations Industrial Development Organization
P.O. Box 300
A-1400 Vienna, Austria
Fax: +43-1-26026-6802
Email: stat@unido.org
www.unido.org/statistics

World Statistics on Mining and Utilities 2018

UNITED NATIONS INDUSTRIAL DEVELOPMENT ORGANIZATION
VIENNA
2018

Published by
Edward Elgar Publishing Limited
The Lypiatts
15 Lansdown Road
Cheltenham
Glos GL50 2JA
UK

Edward Elgar Publishing, Inc.
The William Pratt House
9 Dewey Court
Northampton
Massachusetts 01060-3815
USA

A catalogue record for this book
is available from the British Library

This book is available electronically in the **Elgar**online
Economics subject collection
DOI 10.4337/9781788974585

The description and classification of countries, economies and territories in this publication and the arrangement of the material do not imply the expression of any opinion whatsoever on the part of the Secretariat of the United Nations Industrial Development Organization concerning the legal status of any country, territory, city or area, or of its authorities, or concerning the delimitation of its frontiers or boundaries, or regarding its economic system or degree of development.

The designation "country, economy or area" covers countries, economies, territories, cities and areas. The designation "region, country, economies or area" extends the preceding one by including groupings of entities designated as "country, economy or area". Designations such as "industrialized" and "developing and emerging industrial" economies are intended for statistical convenience and do not necessarily express a judgement about the stage reached by a particular country, economy or area in the development process.

FSC www.fsc.org MIX Paper from responsible sources FSC® C013604

ISBN 978 1 78897 457 8 (cased)
ISBN 978 1 78897 458 5 (ebook)

Printed and bound by CPI Group (UK) Ltd, Croydon, CR0 4YY

ABOUT THIS PUBLICATION

This is the fifth issue of the World Statistics on Mining and Utilities. Following its first release in 2010 the publication was distributed to the National Statistical Offices (NSOs), international organizations and other data users. UNIDO also informed the forty-second session of the United Nations Statistical Commission about the publication with a brief note on its sources and methods and periodicity. UNIDO sincerely appreciates the cooperation with NSOs, OECD (Organisation for Economic Co-operation and Development), UNSD (United Nations Statistics Division) and other international organizations in providing data for this publication.

Wide interest expressed by international data users, especially knowledge institutions and development partners, in mining and utilities statistics confirms the relevance of this publication. Particular interest was shown in statistics on energy related sectors such as production of coal, crude petroleum, natural gas as well as the production and distribution of electricity and gas.

The publication was prepared by a team of statistical staff under the overall supervision of Shyam Upadhyaya, Chief Statistician of UNIDO. Valentin Todorov developed the software system and produced the final digital manuscript. Romana Bauer, Kateryna Gumeniuk, P.Y. Rita Liang and Romana Ransmayr completed the data preparation and the compilation of the country tables. Anja Sedic collected value added data from international sources and prepared the statistical tables for part I. Nelson Correa Fernandez contributed to the final checking of statistical tables.

The team would also like to express its sincere thanks to Cecilia Mercedes Ugaz Estrada, Director of the Department of Policy Research and Statistics of UNIDO, for her support in bringing out this publication.

UNIDO has granted the exclusive publishing rights of the *World Statistics on Mining and Utilities* to Edward Elgar Publishing Limited. The continued partnership with Edward Elgar Publishing Limited on UNIDO statistical publications is greatly appreciated.

List of abbreviations and acronyms:

CIS	Commonwealth of Independent States
ECOWAS	Economic Community of West African States
EU	European Union
GCC	Gulf Cooperation Council (consisting of Bahrain, Kuwait, Oman, Qatar, Saudi Arabia, United Arab Emirates)
GDP	gross domestic product
ISIC	International Standard Industrial Classification of All Economic Activities
MVA	manufacturing value added
n.e.c.	not elsewhere classified
NSOs	National Statistical Offices
OECD	Organisation for Economic Co-operation and Development
OPEC	Organization of the Petroleum Exporting Countries
UNIDO	United Nations Industrial Development Organization
UNSD	United Nations Statistics Division

CONTENTS

Part II. COUNTRY TABLES

DATA AVAILABLE ONLINE AND ON CD-ROM

Database	Description
INDSTAT2: UNIDO Industrial Statistics Database at the 2-digit level of ISIC code (Revision 3)	Long time series data for more than 170 countries/areas compiled on eight principle indicators (number of establishments, total employment, female employment, wages and salaries paid to employees, output, value added, gross fixed capital formation and index numbers of industrial production) starting from 1963 onwards. The data are arranged at the 2-digit level of ISIC (Revision 3) pertaining to the manufacturing sector, which comprises 23 manufacturing divisions.
INDSTAT4: UNIDO Industrial Statistics Database at the 3- and 4-digit levels of ISIC code (Revision 3 & 4)	Time series data for more than 130 (Revision 3)/90 (Revision 4) countries/areas on seven selected data items including number of establishments, total employment, female employment, wages and salaries paid to employees, output, value added and gross fixed capital formation. The data are arranged at the 3- and 4-digit levels of ISIC (Revision 3 & 4) pertaining to the manufacturing sector.
MINSTAT: UNIDO Mining & Utilities Statistics Database at the 2- and 3-digit levels of ISIC code (Revision 3 & 4)	Time series data for more than 110 (Revision 3)/70 (Revision 4) countries/areas on seven selected data items including number of establishments, total employment, female employment, wages and salaries paid to employees, output, value added and gross fixed capital formation. The data are arranged at the 2- and 3-digit levels of ISIC (Revision 3 & 4) pertaining to the mining and utilities sectors.
IDSB: UNIDO Industrial Demand-Supply Balance Database at the 4-digit level of ISIC code (Revision 3 & 4)	Time series data for more than 120 (Revision 3)/90 (Revision 4) countries/ areas on output, imports, exports and apparent consumption (all in current United States dollars) at the 4-digit level of ISIC Revision 3 & 4.
UNIDO Manufacturing Value Added (MVA) Database	Country data on GDP, MVA and population for various regions and groups of economies displayed at constant 2010 and at current United States dollars prices.
UNIDO CIP Database	Country data on ranks and values of Competitive Industrial Performance (CIP) index and its component indicators.

Further details on above products may be obtained from:
Statistics Division
UNIDO
Email: stat@unido.org
Internet: http://www.unido.org/statistics

1 INTRODUCTION

The Statistics on Mining and Utilities are compiled using data collected by UNIDO from the National Statistical Offices as well as those obtained from the Organisation for Economic Co-operation and Development. The publication intends to present the detail country data on major indicators of statistics on mining and utilities at 2-digit and/or 3-digit level of International Standard Industrial Classification of All Economic Activities (ISIC), Revision 3 and Revision 4.

The main purpose of this publication is to provide internationally comparable statistics on mining and utilities sectors on major indicators. Concepts and methods applied to this publication, which are fully compatible to those used in the *International Yearbook of Industrial Statistics*, are briefly described below.

1.1 Scope and coverage

According to the International Recommendations for Industrial Statistics 2008, industrial statistics are those related to the economic activities defined under following sections of ISIC Revision 4:

Section B: Mining and quarrying
Section C: Manufacturing
Section D: Electricity, gas, steam and air conditioning supply
Section E: Water supply; sewerage, waste management and remediation activities

Data pertaining to Section C are published in the *International Yearbook of Industrial Statistics* (latest edition 2018). This publication complements the latter and refers to Sections B, D and E, which are broadly understood as mining and utilities sectors. In ISIC Revision 3, these sectors correspond to following categories:

Category C: Mining and quarrying
Category E: Electricity, gas and water supply

Details of the industry groups under these sections and categories with 2-digit and 3-digit codes are given in appendix II.

The primary source of data at the country level is mainly the annual industrial survey which defines the coverage of data for the reporting country.

1.2 Sources and methods

Part I of the publication represents combined value added data for mining and utilities sectors. These data were obtained from various national and international sources including the United Nations Statistics Division (UNSD). Data from the external sources were supplemented with estimates of UNIDO statisticians.

The country tables presented in part II are compiled from the data directly collected by UNIDO from different national sources, as well as those collected by OECD and provided to UNIDO.

1.3 Part I. Summary tables

Value added of mining and utilities are estimated in accordance with the national accounting concept. In this part, the sum of total value added of Sections B, D and E (Revision 4) and Categories C and E (Revision 3) are presented for different country groups and years. These values indicate the actual contribution of mining and utilities sectors to GDP of a country.

Country groupings, listed in appendix I, are arranged according to the established classification practices in other UNIDO statistical publications.

1.4 Part II. Country tables

Part II presents tables for the major indicators of statistics on mining and utilities, namely: number of establishments, number of employees (or, if not reported, number of persons engaged), wages and salaries paid to employees, output and value added. All value data are presented in current national currencies.

1.4.1 Statistical unit

For most countries and areas, data presented in this publication represent the activity of "establishments" classified to specific industry as recommended in the *International Standard Industrial Classification of All Economic Activities,* Revision 3 and Revision 4. In few cases, statistical unit is "enterprise", "local unit" or "kind of activity unit". An "establishment" is ideally defined as an economic unit that engages, under a single ownership or control, in one, or predominantly one, kind of activity at a single physical location. An "enterprise" is a legal entity possessing the right to conduct business in its own name; for example, to enter into contracts, own property, incur liability for debts, and establish bank accounts. An enterprise may comprise one or more establishments with different locations.

1.4.2 Reference period

The statistics in the tables relate, in general, to the calendar year. It should be noted, however, that in many cases where the basic reference period of the industrial inquiry is the calendar year, returns covering proximate fiscal years may be accepted for reporting purposes and the data for these years incorporated in the calendar-year aggregate without adjustment. In a few countries, fiscal years normally used for public accounting purposes have been adopted as the basic reference periods. In the case of fiscal-year coverage, the year indicated in the tables refers to the calendar year in which the major part of the fiscal year falls. In the case of fiscal years from 1 July to 30 June, the year referred to is normally the one in which the fiscal year ends.

1.4.3 Description of data items

Data items are described in accordance with the International Recommendations for Industrial Statistics. All values are in national currency units and are at current prices unless otherwise indicated.

(1) Number of persons engaged and number of employees

The number of persons engaged is defined as the total number of persons who worked in or for the establishment during the reference year. The concept covers working proprietors, active business partners and unpaid family workers as well as employees, but excludes home workers. The figures reported refer normally to the average number of persons engaged during the reference year, obtained as the sum of the "average number of employees" during the year and the total number of other persons engaged measured for a single period of the year. The category "employees" is intended to include all persons engaged other than working proprietors, active business partners and unpaid family workers. In this publication, preference has been given, whenever possible, to employees over persons engaged.

(2) Wages and salaries

Estimates of wages and salaries include all payments in cash or in kind made to "employees" during the reference year in relation to work done for the establishment. Payments include: (a) direct wages and salaries; (b) remuneration for time not worked; (c) bonuses and gratuities; (d) housing allowances and family allowances paid directly by the employer; and (e) payments in kind.

Compensation of employees is equivalent to wages and salaries plus employers' contributions on behalf of their employees paid to social security, pension and insurance schemes, as well as the benefits received by employees under these schemes and severance and termination pay.

(3) Output

The measure of output normally used in the tables is the census concept that covers only activities of an industrial nature. The value of census output in the case of estimates compiled on a production basis comprises: (a) the value of sale of all products of the establishment; (b) the net change between the beginning and the end of the reference period in the value of work in progress and stocks of goods to be shipped in the same condition as received; (c) the value of industrial work done or industrial services rendered to others; (d) the value of goods shipped in the same condition as received less the amount paid for these goods; and (e) the value of fixed assets produced during the period by the unit for its own use. In the case of estimates compiled on a shipment basis, the net change in the value of stocks of finished goods between the beginning and the end of the reference period is also included.

Gross output is equivalent to census output plus the revenue from activities of a non-industrial nature. Valuation methods differ from country to country. An increasing number of countries are reporting data at basic prices, which excludes taxes on commodity and includes commodity related subsidies. Other methods of reporting valuation are: (a) at factor costs, which excludes all indirect taxes falling on production and includes all current subsidies received in support of production activities; and (b) at producers' prices, which includes all indirect taxes except value added tax or any other deductible taxes and excludes all subsidies.

(4) Value added

The measure of value added normally used in the tables is the census concept, which is defined as the value of census output less the value of census input. Items covered in the latter include: (a) value of materials and supplies for production (including cost of all fuels and electricity purchased); and (b) cost of industrial services received (mainly payments for contract and commission work and repair and maintenance work). If input estimates are compiled on a "received" rather than on a "consumed" basis, the result is adjusted for the net change between the beginning and the end of the period in the value of stocks of materials, fuel and other supplies.

Total value added is the national accounting concept. It is ideally represented by the contribution of the establishments in each branch of activity to the gross domestic product. For the measure of total value added, the cost of non-industrial services is deducted and the receipts for non-industrial services are added to census value added. The estimates, whether in terms of census value added or total value added, are gross of depreciation and other provisions for capital consumption, unless otherwise stated. The valuation may be at factor costs, at basic prices or at producers' prices, depending on the treatment of indirect taxes and subsidies as described above.

Appendix I

LIST OF COUNTRIES AND AREAS INCLUDED IN SELECTED GROUPINGS

Industrialized Economies

EU [a/]

Austria
Belgium
Czechia
Denmark
Estonia
Finland
France
Germany
Hungary
Ireland
Italy
Lithuania
Luxembourg
Malta
Netherlands
Portugal
Slovakia
Slovenia
Spain
Sweden
United Kingdom

East Asia

China, Hong Kong SAR
China, Macao SAR
China, Taiwan Province
Japan
Malaysia
Republic of Korea
Singapore

West Asia

Bahrain
Kuwait
Qatar
United Arab Emirates

Others

Aruba
Australia
British Virgin Islands
Cayman Islands
Curaçao
French Guiana
French Polynesia
Guam
Israel
New Caledonia
New Zealand
Puerto Rico
Trinidad and Tobago
United States Virgin Islands

Other Europe

Andorra
Belarus
Iceland
Liechtenstein
Monaco
Norway
Russian Federation
San Marino
Switzerland

North America

Bermuda
Canada
Greenland
United States of America

a/ Excluding non-industrialized EU economies.

Appendix I

(continued)

LIST OF COUNTRIES AND AREAS INCLUDED IN SELECTED GROUPINGS

Developing and Emerging Industrial Economies by Region

AFRICA

Central Africa

Cameroon
Central African Republic
Chad
Congo
Equatorial Guinea
Gabon
Sao Tome and Principe

Eastern Africa

Burundi
Comoros
Djibouti
Eritrea
Ethiopia
Kenya
Réunion
Rwanda
Somalia
Uganda

North Africa

Algeria
Egypt
Libya
Morocco
South Sudan
Sudan
Tunisia

Southern Africa

Angola
Botswana
Democratic Rep of the Congo
Lesotho
Madagascar
Malawi
Mauritius
Mozambique
Namibia
Seychelles
South Africa
Swaziland
United Republic of Tanzania
Zambia
Zimbabwe

Western Africa

Benin
Burkina Faso
Cabo Verde
Côte d'Ivoire
Gambia
Ghana
Guinea
Guinea-Bissau
Liberia
Mali
Mauritania
Niger
Nigeria
Senegal
Sierra Leone
Togo

ASIA & PACIFIC

China

Central Asia

Kazakhstan
Kyrgyzstan
Mongolia
Tajikistan
Turkmenistan
Uzbekistan

South Asia

Afghanistan
Bangladesh
Bhutan
India
Maldives
Nepal
Pakistan
Sri Lanka

South East Asia

Brunei Darussalam
Cambodia
Indonesia
Lao People's Dem Rep
Myanmar
Philippines
Thailand
Viet Nam

West Asia

Armenia
Azerbaijan
Iran (Islamic Republic of)
Iraq
Jordan
Lebanon
Oman
Saudi Arabia
State of Palestine
Syrian Arab Republic
Yemen

Other Asia & Pacific

Cook Islands
Democratic People's Rep of Korea
Fiji
Kiribati
Marshall Islands
Micronesia, Federated States of
Palau
Papua New Guinea
Samoa
Solomon Islands
Timor-Leste
Tonga
Tuvalu
Vanuatu

EUROPE

Albania
Bosnia and Herzegovina
Bulgaria
Croatia
Cyprus
Georgia
Greece
Latvia
Montenegro
Poland
Republic of Moldova
Romania
Serbia
The f. Yugosl. Rep of Macedonia
Turkey
Ukraine

LATIN AMERICA

Caribbean

Anguilla
Antigua and Barbuda
Bahamas
Barbados
Cuba
Dominica
Dominican Republic
Grenada
Guadeloupe
Haiti
Jamaica
Martinique
Montserrat
Saint Kitts and Nevis
Saint Lucia
Saint Vincent and the Grenadines

Central America

Belize
Costa Rica
El Salvador
Guatemala
Honduras
Mexico
Nicaragua
Panama

South America

Argentina
Bolivia (Plurinational State of)
Brazil
Chile
Colombia
Ecuador
Guyana
Paraguay
Peru
Suriname
Uruguay
Venezuela (Bolivarian Republic of)

Appendix II

Detailed description of International Standard Industrial Classification of All Economic Activities (ISIC) - Revision 3

ISIC	Description
C	**MINING AND QUARRYING**
Division 10	**Mining of coal and lignite; extraction of peat**
101	Mining and agglomeration of hard coal
102	Mining and agglomeration of lignite
103	Extraction and agglomeration of peat
Division 11	**Extraction of crude petroleum and natural gas; service activities incidental to oil and gas extraction, excluding surveying**
111	Extraction of crude petroleum and natural gas
112	Service activities incidental to oil and gas extraction excluding surveying
Division 12	**Mining of uranium and thorium ores**
120	Mining of uranium and thorium ores
Division 13	**Mining of metal ores**
131	Mining of iron ores
132	Mining of non-ferrous metal ores, except uranium and thorium ores
Division 14	**Other mining and quarrying**
141	Quarrying of stone, sand and clay
142	Mining and quarrying n.e.c.
E	**ELECTRICITY, GAS AND WATER SUPPLY**
Division 40	**Electricity, gas, steam and hot water supply**
401	Production, collection and distribution of electricity
402	Manufacture of gas; distribution of gaseous fuels through mains
403	Steam and hot water supply
Division 41	**Collection, purification and distribution of water**
410	Collection, purification and distribution of water

Appendix II (continued)

Detailed description of International Standard Industrial Classification of All Economic Activities (ISIC) - Revision 4

ISIC	Description
B	**MINING AND QUARRYING**
Division 05	**Mining of coal and lignite**
051	Mining of hard coal
052	Mining of lignite
Division 06	**Extraction of crude petroleum and natural gas**
061	Extraction of crude petroleum
062	Extraction of natural gas
Division 07	**Mining of metal ores**
071	Mining of iron ores
072	Mining of non-ferrous metal ores
Division 08	**Other mining and quarrying**
081	Quarrying of stone, sand and clay
089	Mining and quarrying n.e.c.
Division 09	**Mining support service activities**
091	Support activities for petroleum, natural gas extraction
099	Support activities for other mining and quarrying
D	**ELECTRICITY, GAS, STEAM AND AIR CONDITIONING SUPPLY**
Division 35	**Electricity, gas, steam and air conditioning supply**
351	Electric power generation, transmission and distribution
352	Manufacture of gas; distribution of gaseous fuels through mains
353	Steam and air conditioning supply
E	**WATER SUPPLY; SEWERAGE, WASTE MANAGEMENT AND REMEDIATION ACTIVITIES**
Division 36	**Water collection, treatment and supply**
360	Water collection, treatment and supply
Division 37	**Sewerage**
370	Sewerage
Division 38	**Waste collection, treatment and disposal activities; materials recovery**
381	Waste collection
382	Waste treatment and disposal
383	Materials recovery
Division 39	**Remediation activities and other waste management services**
390	Remediation activities and other waste management services

Part I

SUMMARY TABLES

An overview of the world's mining and utilities

The mining and utilities sectors comprise mining and quarrying activities as well as the supply of electricity, gas and water, waste collection, disposal and remediation activities, which are broadly referred to as utilities. Mining and quarrying activities produce essential raw materials for further processing by the manufacturing industry, while the products and services of the utilities sector are directly supplied to consumers via a permanent infrastructure of lines, mains and pipes. The scope of the utilities sector has been expanded in Revision 4 of the International Standard Industrial Classification of All Economic Activities to include the supply of hot water, steam and air-conditioning. The production activities of mining and quarrying include the extraction of minerals that occur naturally such as solids (coal and ores), liquids (petroleum) or gases (natural gas), as well as quarrying of stone, sand, clay and various other materials. The current edition of World Statistics on Mining and Utilities highlights the growth and structure of these sectors at country, regional and global level.

Statistics on mining and utilities have acquired new relevance in recent years. The forty-ninth session of the United Nations Statistical Commission endorsed[1] a new *Handbook for Compiling Statistics for Economies Based on Natural Resources* as international guidelines for the production of statistics covering mining and related activities. Material consumption has also been an important component of sustainable development goals (SDG), especially those related to sustainable production and consumption. The current edition of World Statistics on Mining and Utilities is largely in compliance with the guidelines of the above-mentioned Handbook, and the statistics presented here might be relevant for monitoring the progress made by countries towards achieving the SDGs. The overall trend observed in the mining and utilities sectors is influenced by increasing consumption of utility products and services. However, the share of mining in GDP remains dominant in many countries. The value added growth of mining activities is generally low, which is attributed to two main reasons. First, the gradual depletion of minerals resources has reduced the growth potential of mineral products in many countries. Secondly, lower crude oil prices have significantly influenced the production value of oil, which accounts for a significant share of mining and quarrying value added in several countries.

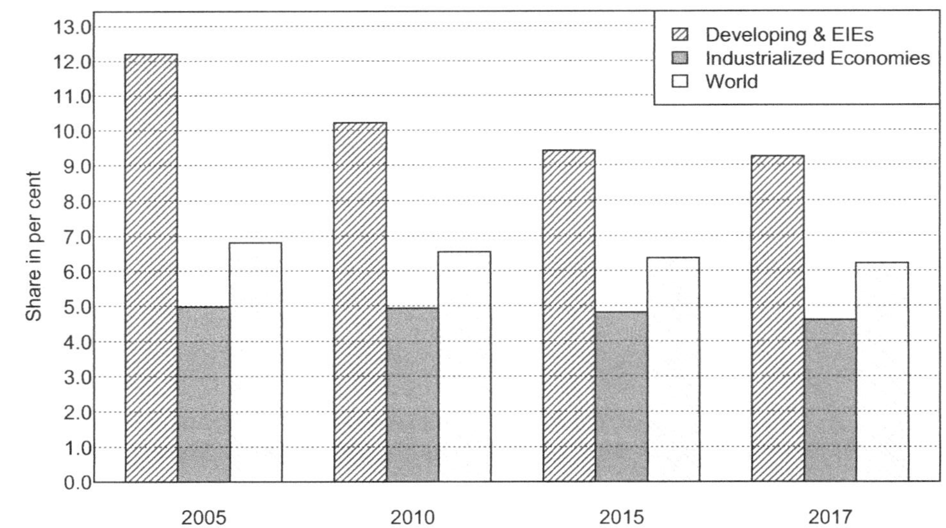

Source: UNIDO Database

Figure 1: Share of value added of the mining and utilities sectors in GDP at constant 2010 USD

Figure 1 illustrates the share of value added of the mining and utilities sectors in GDP over the period 2005–2017. The figure suggests that the net output produced in these sectors has grown at a slower rate in comparison with other activities in the economy. Thus, we observe a slightly downward trend at world level, with this share dropping from 6.8 per cent in 2005 to 6.2 per cent in 2017. This decreasing trend was registered both in industrialized and in developing economies, which confirms the global nature of the decrease in the share of mining and utilities in the economy. Figure 1 also shows a considerable difference in the size of the share of these activities between the two country groups. The contribution of the mining and utilities sectors to GDP is much higher in developing countries than in industrialized economies. This discrepancy is due to the difference in distribution of natural resources each country has, and more significantly, to the fact that the economy of developing countries is traditionally more natural resource-based than industrialized economies.

[1] UN Statistical Commission Document E/CN.3/2018/12 - https://unstats.un.org/unsd/statcom/49th-session/documents/2018-12-UlaanbaatarGroup-E.pdf

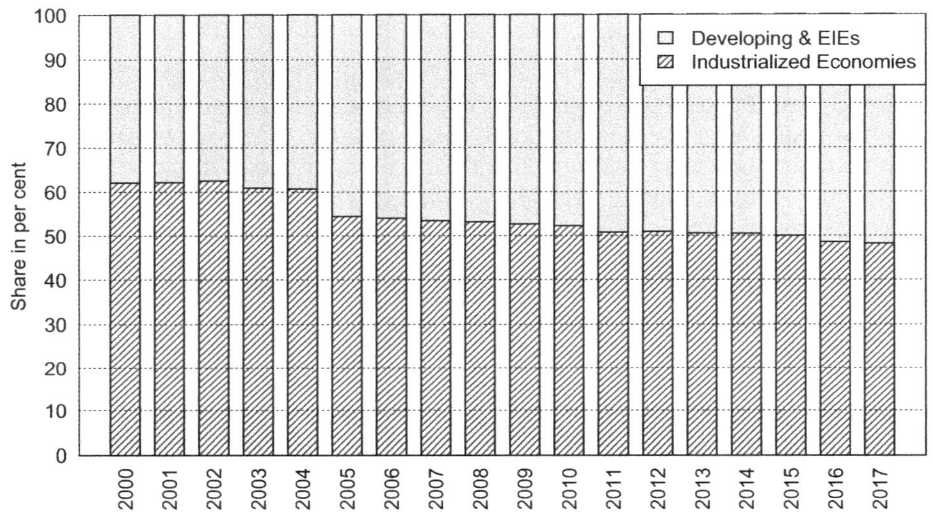

Source: UNIDO Database

Figure 2: Distribution of world value added of mining and utilities activities by country group at constant 2010 USD

While production in the mining and quarrying sector depends on the availability of natural resources in the country, utilities can be made available in every country. The per capita consumption of electricity, water and other utility services is relatively high in industrialized economies compared to developing economies. However, the opposite applies to production in mining and quarrying, as its contribution to GDP is relatively higher in developing economies. Figure 2 reflects these trends, revealing a decreasing share of industrialized countries in total world output of mining and utilities activities. Industrialized countries have reduced their share to the world total output of mining and utilities; they accounted for nearly 62 per cent in the 2000, but only for a 48 per cent in 2017. The opposite trend took place in developing countries as they went from 38 per cent to 52 per cent during the same period. This increase can be also explained by the rise in demand for energy products such as electricity, natural gas and coal to boost their manufacturing sectors' growth rate.

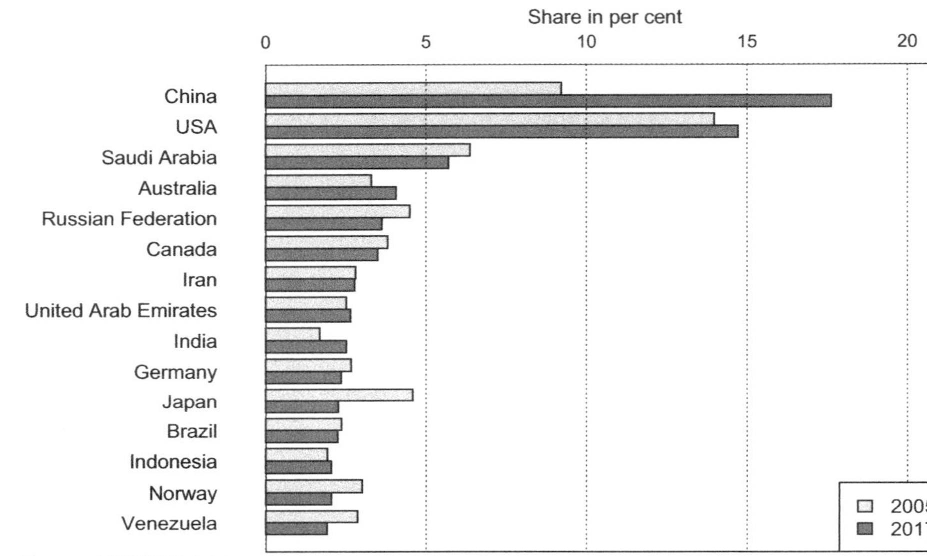

Source: UNIDO Database

Figure 3: Share of major economies in world total value added of mining and utilities activities at constant 2010 USD

Figure 3 reveals that the share of mining and utilities sectors in world production has decreased since 2005 and that this trend was followed by both developed and developing economies. A more disaggregated analysis shows significant heterogeneities across countries and suggests that a concentration process has taken place in favour of the two major world producers. The recent policy of the U.S. government is aimed at stimulating the domestic production of mineral resources and has resulted in an increase of output from mining activities. The increase of the share of world value added was much higher in China, which made China the largest producer of mining and utilities in the world. By contrast, the total value added of mining and utilities dropped considerably, as shown in Figure 3, in many other important economies such as Canada, Germany, Japan, Norway and Saudi Arabia.

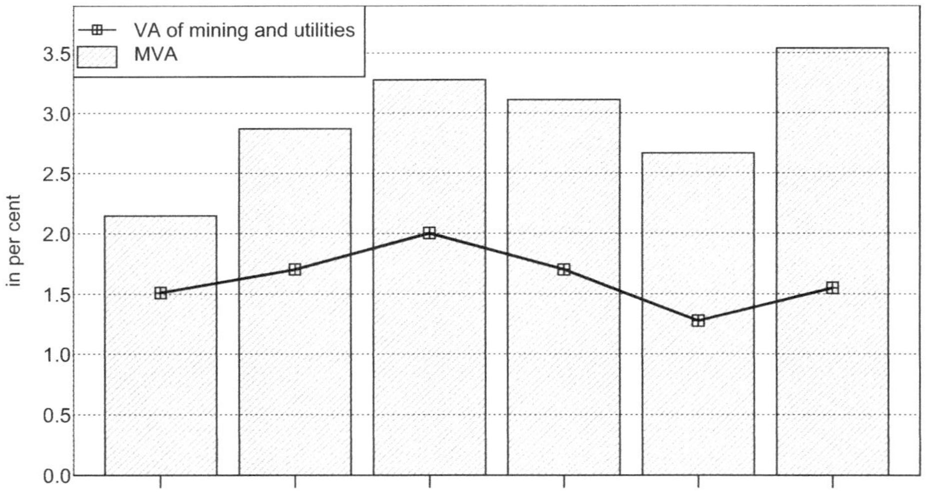

Source: UNIDO Database

Figure 4: Annual growth rate of global value added of mining and utilities and of manufacturing at constant 2010 USD

Demand for mining and utility products is very high. While demand for mining products from production sectors, such as manufacturing and construction, is high, demand for utilities is mostly associated with household consumption. This means that demand for utilities remains relatively stable and is less affected by economic fluctuations during periods of production boom or recessions. Figure 4 presents the growth of value added in manufacturing and compares it with that of value added in mining and utilities for the period 2012 to 2017, that is, after the financial crisis and the consequent period of recovery from 2009 to 2011. Figure 4 shows that both sectors experienced a positive period of growth, albeit with some differences. While world manufacturing value added capitalized on the favourable conditions of the world economy and ranged between 2 to 4 per cent annually, the value added of mining and utilities grew at a much slower yet stable rate of between 1 and 2 per cent.

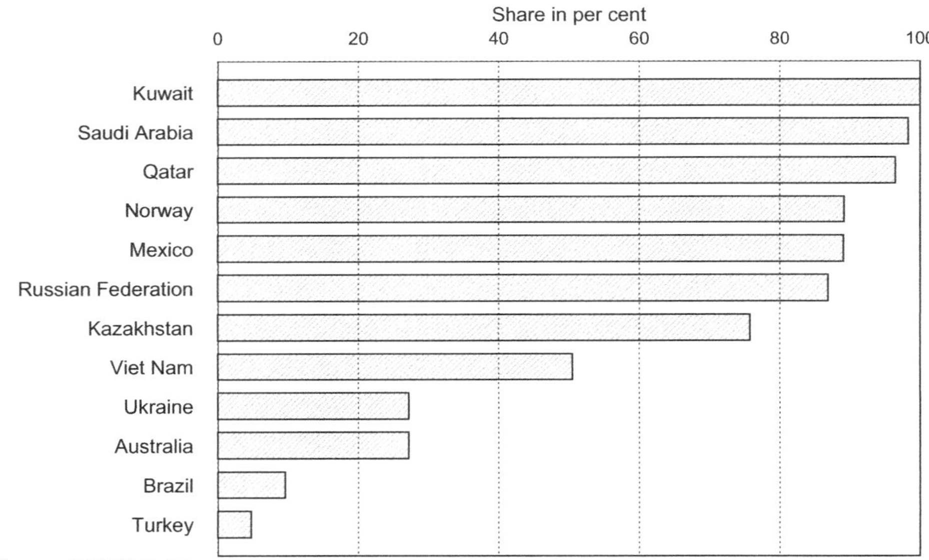

Source: UNIDO Database

Figure 5: Share of extraction of crude petroleum and natural gas in total mining and quarrying for selected economies at current prices, 2015 or latest year available

The extraction of crude petroleum and natural gas is one of the most important mining activities supplying essential minerals to meet the energy demand of both industries and households. Yet, the share of this industry in the total value added of the mining and quarrying sector varies considerably across countries, as their availability of natural resources differs greatly. Figure 5 presents the value of this share for 12 selected economies. This share is very high for several Middle Eastern countries, such as Kuwait, Saudi Arabia and Qatar, as they have registered values above 90%. By contrast, other countries with a relatively higher share of mining activities in their economies such as Australia, Brazil, Turkey and Ukraine, have a much smaller contribution of crude petroleum and natural gas to their total value added of mining and quarrying. This comparison highlights the immense heterogeneity that exists in countries' natural resource endowments.

Table 1.1 **DISTRIBUTION OF WORLD VALUE ADDED OF MINING AND UTILITIES SECTORS, 2005-2017**

Year	Industrialized Economies					Developing and Emerging Industrial Economies			
	Europe	East Asia	West Asia	North America	Others	Africa	Asia & Pacific	Europe	Latin America
	Percentage share in world total value added (at constant 2010 prices)								
2005	21.3	6.2	5.4	17.8	3.7	6.4	25.7	2.4	11.1
2006	20.1	6.1	5.7	18.3	3.8	6.3	26.1	2.5	11.1
2007	19.7	5.8	5.5	18.6	3.9	6.3	26.8	2.4	11.0
2008	19.5	5.9	5.6	18.4	3.8	6.1	27.4	2.4	10.9
2009	19.3	5.2	5.3	18.8	4.1	6.1	28.0	2.3	10.9
2010	18.8	5.3	5.5	18.6	4.1	6.2	28.4	2.3	10.8
2011	17.6	4.7	6.0	18.5	4.0	6.1	30.0	2.4	10.7
2012	17.7	4.0	6.3	18.8	4.1	6.2	29.9	2.3	10.7
2013	16.9	4.2	6.2	19.1	4.2	6.0	30.5	2.3	10.6
2014	16.4	4.2	6.1	19.4	4.4	5.8	31.4	2.2	10.1
2015	16.2	3.8	6.2	19.4	4.4	5.7	32.2	2.2	9.9
2016 a/	16.1	3.6	6.2	18.2	4.4	5.5	34.0	2.2	9.8
2017 b/	15.9	3.7	5.9	18.2	4.5	5.4	34.6	2.2	9.6

a/ Provisional.
b/ Estimate.

Table 1.2 DISTRIBUTION OF WORLD VALUE ADDED OF MINING AND UTILITIES SECTORS AND POPULATION, SELECTED YEARS
(Percentage)

Country group	Value added at constant 2010 prices					Population				
	2005	2010	2015	2016 [a/]	2017 [b/]	2005	2010	2015	2016	2017
Africa	6.4	6.2	5.7	5.6	5.5	12.6	13.5	14.4	14.7	14.2
Asia & Pacific	40.8	43.1	46.5	48.1	48.5	60.9	60.6	60.3	60.2	60.6
Europe	23.8	21.2	18.4	18.3	18.1	12.6	12.0	11.5	11.4	11.4
Latin America	11.2	10.9	10.0	9.8	9.7	8.8	8.8	8.8	8.8	8.8
North America	17.8	18.6	19.4	18.2	18.2	5.1	5.1	5.0	4.9	5.0
World	100.0	100.0	100.0	100.0	100.0	100.0	100.0	100.0	100.0	100.0

a/ Provisional.
b/ Estimate.

Table 1.3 **GROWTH OF TOTAL AND PER-CAPITA VALUE ADDED OF MINING AND UTILITIES SECTORS, AT CONSTANT 2010 PRICES**

Country groups	Total value added		Per-capita value added	
	2005-2010	2010-2017	2005-2010	2010-2017
Industrialized Economies	0.6	0.8	-0.1	0.3
EU [a/]	-0.3	-0.7	-0.8	-0.9
Other Europe	-1.4	-0.1	-1.4	-0.3
East Asia	-2.1	-2.9	-2.5	-3.2
West Asia	1.0	2.4	-8.9	-0.5
North America	2.2	1.7	1.3	0.9
Others	3.3	3.6	1.6	2.3
Developing & Emerging Industrial Economies	2.4	2.9	1.0	1.6
Africa	0.6	-0.2	-1.9	-2.3
Asia & Pacific	3.6	4.6	2.4	3.5
Europe	0.2	0.2	0.1	-0.1
Latin America	0.8	0.0	-0.4	-1.1
World	1.4	1.8	0.2	0.7

a/ Excluding non-industrialized EU economies.

Table 1.4

SHARE OF VALUE ADDED OF MINING AND UTILITIES SECTORS IN GDP, SELECTED YEARS
(Percentage)

	At constant 2010 prices					
Country groups	**2005**	**2010**	**2014**	**2015**	**2016 b/**	**2017 c/**
Industrialized Economies	5.0	4.9	4.9	4.8	4.7	4.6
EU a/	3.5	3.3	2.9	2.9	2.9	2.8
Other Europe	13.6	11.1	10.1	10.4	10.3	10.2
East Asia	3.3	2.9	2.3	2.1	2.0	2.0
West Asia	49.0	42.9	42.6	42.0	41.5	40.0
North America	4.4	4.8	5.0	5.0	4.7	4.7
Others	9.1	9.5	10.1	10.3	10.0	10.1
Developing & Emerging Industrial Economies	12.2	10.2	9.5	9.4	9.5	9.2
Africa	19.2	15.8	13.9	13.4	13.0	12.3
Asia & Pacific	13.9	11.0	10.2	10.1	10.1	9.9
Europe	5.4	4.9	4.6	4.4	4.5	4.3
Latin America	10.2	8.9	8.3	8.3	8.4	8.3
World	6.8	6.5	6.4	6.4	6.3	6.2

a/ Excluding non-industrialized EU economies.
b/ Provisional.
c/ Estimate.

Part II
COUNTRY TABLES

Albania

ISIC Revision 4		Number of establishments					Number of employees					Wages and salaries paid to employees				
			(number)					(number)					(millions of Albanian Leks)			
ISIC	Industry	Note	2012	2013	2014	2015	Note	2012	2013	2014	2015	Note	2012	2013	2014	2015
05	Mining of coal and lignite		17a/	20a/	27a/	...		4043a/	4076a/	3751a/	...		4066a/	4641a/	5182a/	...
051	Mining of hard coal		...	...	...	...		...	...	...	...		...	...	...	...
052	Mining of lignite		...	...	...	...		...	...	...	...		...	...	...	...
06	Extraction of crude petroleum & natural gas		...a/	...a/	...a/	...		...a/	...a/	...a/	...		...a/	...a/	...a/	...
061	Extraction of crude petroleum		...	...	...	...		...	...	...	...		...	...	...	...
062	Extraction of natural gas		...	...	...	...		...	...	...	...		...	...	...	...
07	Mining of metal ores		497b/	508b/	592b/	...		5461b/	6279b/	7057b/	...		5707b/	6393b/	7089b/	...
071	Mining of iron ores		...	...	...	...		...	...	...	...		...	...	...	...
072	Mining of non-ferrous metal ores		...	...	...	...		...	...	...	...		...	...	...	...
08	Other mining and quarrying		...b/	...b/	...b/	...		...b/	...b/	...b/	...		...b/	...b/	...b/	...
081	Quarrying of stone, sand and clay		...	...	...	...		...	...	...	...		...	...	...	...
089	Mining and quarrying n.e.c.		...	...	...	...		...	...	...	...		...	...	...	...
09	Mining support service activities		...b/	...b/	...b/	...		...b/	...b/	...b/	...		...b/	...b/	...b/	...
091	Support activities for petroleum		...	...	...	...		...	...	...	...		...	...	...	...
099	Support activities for other mining, quarrying		...	...	...	...		...	...	...	...		...	...	...	...
B	Mining and quarrying		513	528	619	...		9504	10356	10807	...		9773	11033	12271	...
35	Electricity, gas, steam & air conditioning		160	258	237	...		6890	7216	8103	...		5213	5073	5853	...
351	Electric power generation, transmission		...	...	...	...		...	...	...	...		...	...	...	...
352	Manufacture of gas		...	...	...	...		...	...	...	...		...	...	...	...
353	Steam and air conditioning supply		...	...	...	...		...	...	...	...		...	...	...	...
D	Electricity, gas, steam & air conditioning		160	258	237	...		6890	7216	8103	...		5213	5073	5853	...
36	Water collection, treatment and supply		61	60	79	...		5513	5715	5852	...		2154	2271	2337	...
37	Sewerage		188c/	224c/	215c/	...		4027c/	4447c/	4455c/	...		1471c/	1494c/	1603c/	...
38	Waste collection,treatment,disposal activities		...c/	...c/	...c/	...		...c/	...c/	...c/	...		...c/	...c/	...c/	...
381	Waste collection		...	...	...	...		...	...	...	...		...	...	...	...
382	Waste treatment and disposal		...	...	...	...		...	...	...	...		...	...	...	...
383	Materials recovery		...	...	...	...		...	...	...	...		...	...	...	...
39	Remediation activities		...c/	...c/	...c/	...		...c/	...c/	...c/	...		...c/	...c/	...c/	...
E	Water supply;sewerage,waste management		249	284	294	...		9540	10162	10307	...		3625	3766	3940	...

a/ 05 includes 06.
b/ 07 includes 08 and 09.
c/ 37 includes 38 and 39.

Albania

ISIC Revision 4		Output at basic prices					Value added at basic prices				
			(millions of Albanian Leks)					(millions of Albanian Leks)			
ISIC	Industry	Note	2012	2013	2014	2015	Note	2012	2013	2014	2015
05	Mining of coal and lignite		43756a/	55922a/	64693a/	...		32145a/	38162a/	45009a/	...
051	Mining of hard coal		...	...	...	...		...	...	...	...
052	Mining of lignite		...	...	...	...		...	...	...	...
06	Extraction of crude petroleum & natural gas		...a/	...a/	...a/	...		...a/	...a/	...a/	...
061	Extraction of crude petroleum		...	...	...	...		...	...	...	...
062	Extraction of natural gas		...	...	...	...		...	...	...	...
07	Mining of metal ores		31647b/	35557b/	40936b/	...		17431b/	18610b/	19807b/	...
071	Mining of iron ores		...	...	...	...		...	...	...	...
072	Mining of non-ferrous metal ores		...	...	...	...		...	...	...	...
08	Other mining and quarrying		...b/	...b/	...b/	...		...b/	...b/	...b/	...
081	Quarrying of stone, sand and clay		...	...	...	...		...	...	...	...
089	Mining and quarrying n.e.c.		...	...	...	...		...	...	...	...
09	Mining support service activities		...b/	...b/	...b/	...		...b/	...b/	...b/	...
091	Support activities for petroleum		...	...	...	...		...	...	...	...
099	Support activities for other mining, quarrying		...	...	...	...		...	...	...	...
B	Mining and quarrying		75403	91479	105629	...		49576	56772	64816	...
35	Electricity, gas, steam & air conditioning		76940	89526	88603	...		13510	25101	24966	...
351	Electric power generation, transmission		...	...	...	...		...	...	...	...
352	Manufacture of gas		...	...	...	...		...	...	...	...
353	Steam and air conditioning supply		...	...	...	...		...	...	...	...
D	Electricity, gas, steam & air conditioning		76940	89526	88603	...		13510	25101	24966	...
36	Water collection, treatment and supply		7823	7214	7268	...		4955	4249	4335	...
37	Sewerage		9194c/	7035c/	8729c/	...		2575c/	2927c/	2760c/	...
38	Waste collection,treatment,disposal activities		...c/	...c/	...c/	...		...c/	...c/	...c/	...
381	Waste collection		...	...	...	...		...	...	...	...
382	Waste treatment and disposal		...	...	...	...		...	...	...	...
383	Materials recovery		...	...	...	...		...	...	...	...
39	Remediation activities		...c/	...c/	...c/	...		...c/	...c/	...c/	...
E	Water supply;sewerage,waste management		17017	14249	15997	...		7530	7177	7096	...

a/ 05 includes 06.
b/ 07 includes 08 and 09.
c/ 37 includes 38 and 39.

Algeria

ISIC Revision 4		Number of establishments					Number of employees					Wages and salaries				
			(number)					(number)					(Algerian Dinars)			
ISIC	Industry	Note	2012	2013	2014	2015	Note	2012	2013	2014	2015	Note	2012	2013	2014	2015
05	Mining of coal and lignite		6	7	7	5		103	103	103	103		...	...	...	...
06	Extraction of crude petroleum & natural gas		57	57	58	50		66	66	66	66		...	...	...	...
07	Mining of metal ores		27	27	28	26		52	52	55	55		...	...	...	...
08	Other mining and quarrying		1218	1236	1277	1189		...	...	...	...		...	...	...	...
09	Mining support service activities		397	411	444	433		...	...	...	...		...	...	...	...
B	Mining and quarrying		1705	1738	1814	1703	a/	924	943	968	988		...	...	...	...
35	Electricity, gas, steam & air conditioning		293	319	361	378		888	893	899	908		...	...	...	...
D	Electricity, gas, steam & air conditioning		293	319	361	378		888	893	899	908		...	...	...	...
36	Water collection, treatment and supply		236	250	275	303		...	...	...	...		...	...	...	...
37	Sewerage		82	83	84	80		...	...	...	...		...	...	...	...
38	Waste collection,treatment,disposal activities		500	560	674	685		1781	1860	1992	2135		...	...	...	...
39	Remediation activities		-	-	-	-		-	-	-	-		...	...	...	...
E	Water supply;sewerage,waste management		818	893	1033	1068	a/	1834	1915	2054	2209		...	...	...	...

a/ Sum of available data.

Angola

ISIC Revision 4		Number of establishments					Number of persons engaged					Wages and salaries				
		Note	(number)				Note	(number)				Note	(millions of Angolan Kwanzas)			
ISIC	Industry		2012	2013	2014	2015		2012	2013	2014	2015		2012	2013	2014	2015
051	Mining of hard coal		...	...	...	...		...	...	...	...		...	...	...	...
052	Mining of lignite		...	...	...	...		...	...	...	...		...	...	...	...
061	Extraction of crude petroleum		...	4	4	4		...	3875	3932	4055		...	...	...	...
062	Extraction of natural gas		...	1	1	1		...	205	225	179		...	...	...	...
071	Mining of iron ores		...	...	...	...		...	...	...	...		...	...	...	...
072	Mining of non-ferrous metal ores		...	...	...	...		...	...	...	...		...	...	...	...
081	Quarrying of stone, sand and clay		...	9	11	13		...	275	287	389		...	...	...	...
089	Mining and quarrying n.e.c.		...	...	...	...		...	...	...	...		...	...	...	...
091	Support activities for petroleum		...	2	2	1		...	39	51	26		...	...	...	...
099	Support activities for other mining, quarrying		...	...	...	...		...	...	...	...		...	...	...	...
B	Mining and quarrying		...	21	23	24		...	8474	8652	8883		...	...	...	...
351	Electric power generation, transmission		...	1	1	1		...	1923	1935	2015		...	...	...	...
352	Manufacture of gas		...	...	...	...		...	...	...	...		...	...	...	...
353	Steam and air conditioning supply		...	...	...	...		...	...	...	...		...	...	...	...
D	Electricity, gas, steam & air conditioning		...	1	1	1		...	1923	1935	2015		...	...	...	...
360	Water collection, treatment and supply		...	3	3	3		...	1813	1870	1904		...	...	...	...
370	Sewerage		...	...	...	...		...	...	...	...		...	...	...	...
381	Waste collection		...	...	...	...		...	...	...	...		...	...	...	...
382	Waste treatment and disposal		...	...	...	...		...	...	...	...		...	...	...	...
383	Materials recovery		...	...	...	...		...	...	...	...		...	...	...	...
390	Remediation activities		...	...	...	...		...	...	...	...		...	...	...	...
E	Water supply;sewerage,waste management		...	3	3	3		...	1813	1870	1904		...	...	...	...

Angola

ISIC Revision 4		Output (valuation not defined)					Value added				
			(millions of Angolan Kwanzas)					(millions of Angolan Kwanzas)			
ISIC	Industry	Note	2012	2013	2014	2015	Note	2012	2013	2014	2015
051	Mining of hard coal		...	...	...	...		...	...	...	...
052	Mining of lignite		...	...	...	...		...	...	...	...
061	Extraction of crude petroleum		...	5608285	5781737	5926686		...	...	...	...
062	Extraction of natural gas		...	...	...	...		...	...	...	...
071	Mining of iron ores		...	...	...	...		...	...	...	...
072	Mining of non-ferrous metal ores		...	...	...	...		...	...	...	...
081	Quarrying of stone, sand and clay		...	2136	2810	2860		...	...	...	...
089	Mining and quarrying n.e.c.		...	...	...	...		...	...	...	...
091	Support activities for petroleum		...	299	393	400		...	...	...	...
099	Support activities for other mining, quarrying		...	...	...	...		...	...	...	...
B	Mining and quarrying		...	5611019	5785335	5930346		...	...	...	...
351	Electric power generation, transmission		...	115841	117011	119945		...	...	...	...
352	Manufacture of gas		...	...	...	...		...	...	...	...
353	Steam and air conditioning supply		...	...	...	...		...	...	...	...
D	Electricity, gas, steam & air conditioning		...	115841	117011	119945		...	...	...	...
360	Water collection, treatment and supply		...	12503	13024	12078		...	...	...	...
370	Sewerage		...	...	...	...		...	...	...	...
381	Waste collection		...	...	...	...		...	...	...	...
382	Waste treatment and disposal		...	...	...	...		...	...	...	...
383	Materials recovery		...	...	...	...		...	...	...	...
390	Remediation activities		...	...	...	...		...	...	...	...
E	Water supply;sewerage,waste management		...	12503	13024	12078		...	...	...	...

Armenia

ISIC Revision 4		Number of establishments					Number of employees					Wages and salaries paid to employees				
		Note	(number)				Note	(number)				Note	(millions of Armenian Drams)			
ISIC	Industry		2012	2013	2014	2015		2012	2013	2014	2015		2012	2013	2014	2015
051	Mining of hard coal		...	...	...	...		...	...	...	...		...	...	...	...
052	Mining of lignite		...	...	...	...		...	...	...	...		...	...	...	...
061	Extraction of crude petroleum		...	...	...	...		...	...	...	...		...	...	...	...
062	Extraction of natural gas		...	...	...	...		...	...	...	...		...	...	...	...
071	Mining of iron ores		...	...	...	...		...	...	...	...		...	...	...	...
072	Mining of non-ferrous metal ores		8	9	9	9		...	6965	7057	8332		...	...	...	...
081	Quarrying of stone, sand and clay		51	55	52	53		...	875	784	721		...	...	...	...
089	Mining and quarrying n.e.c.		12	11	10	10		...	134	87	96		...	...	...	...
091	Support activities for petroleum		...	...	...	...		...	...	...	...		...	...	...	...
099	Support activities for other mining, quarrying		...	...	...	...		...	...	...	...		...	...	...	...
B	Mining and quarrying		71	75	71	72		7479	7974	7928	9149		25819	30337	31487	39631
351	Electric power generation, transmission		142	162	170	183		...	13894	13727	13570		...	...	...	...
352	Manufacture of gas		1	1	1	1		...	5502	5444	5490		...	...	...	...
353	Steam and air conditioning supply		1	2	1	-		...	6	4	-		...	...	...	...
D	Electricity, gas, steam & air conditioning		144	165	172	184		20256	19402	19175	19060		43377	45030	51827	51665
360	Water collection, treatment and supply		3	4	4	5		...	2570	2343	2299		...	...	...	...
370	Sewerage		9	7	7	7		...	529	483	480		...	...	...	...
381	Waste collection		66	66	70	66		...	2203	2191	2162		...	...	...	...
382	Waste treatment and disposal		3	3	3	2		...	14	13	12		...	...	...	...
383	Materials recovery		1	1	...	...		...	4	...	...		...	...	...	...
390	Remediation activities		...	...	...	...		...	...	...	...		...	...	...	...
E	Water supply;sewerage,waste management		82	81	84	80		5109	5320	5030	4953		8132	8915	9239	10106

Armenia

ISIC Revision 4		Output at basic prices					Value added at basic prices				
			(millions of Armenian Drams)					(millions of Armenian Drams)			
ISIC	Industry	Note	2012	2013	2014	2015	Note	2012	2013	2014	2015
051	Mining of hard coal		...	...	...	...		...	...	...	...
052	Mining of lignite		...	...	...	...		...	...	...	...
061	Extraction of crude petroleum		...	...	...	...		...	...	...	...
062	Extraction of natural gas		...	...	...	...		...	...	...	...
071	Mining of iron ores		...	...	...	...		...	...	...	...
072	Mining of non-ferrous metal ores		187426	191339	188238	215513		...	...	...	85809
081	Quarrying of stone, sand and clay		4561	4946	4453	4361		...	...	...	2281
089	Mining and quarrying n.e.c.		1339	494	452	793		...	...	...	401
091	Support activities for petroleum		...	...	...	...		...	...	...	...
099	Support activities for other mining, quarrying		...	...	...	...		...	...	...	...
B	Mining and quarrying		193326	196779	193142	220667		...	...	...	88491
351	Electric power generation, transmission		182789	214478	186568	211868		...	...	...	135172
352	Manufacture of gas		28932	36013	45187	50012		...	...	...	28507
353	Steam and air conditioning supply		3	12	1	-		...	...	...	-
D	Electricity, gas, steam & air conditioning		211724	250503	231756	261879		...	...	...	163678
360	Water collection, treatment and supply		11379	12020	12561	13429		...	...	...	9793
370	Sewerage		2682	2089	2162	2216		...	...	...	1287
381	Waste collection		3242	3586	3819	4976		...	...	...	2329
382	Waste treatment and disposal		85	76	84	59		...	...	...	13
383	Materials recovery		50	...	...	...		...	...	...	...
390	Remediation activities		...	...	...	...		...	...	...	...
E	Water supply;sewerage,waste management		17438	17771	18626	20680		...	...	...	13422

Australia

ISIC Revision 4		Number of enterprises					Number of persons engaged					Wages and salaries paid to employees				
			(number)					(number)					(millions of Australian Dollars)			
ISIC	Industry	Note	2012	2013	2014	2015	Note	2012	2013	2014	2015	Note	2012	2013	2014	2015
051	Mining of hard coal		643	...	...	...	a/	45000	44000	45000	41000	a/	6141	6279	6139	5891
052	Mining of lignite		148	...	...	...	a/	...	...	...	...	a/	...	...	...	...
061	Extraction of crude petroleum		129	...	...	...	b/	17000	19000	23000	23000	b/	3391	3717	4494	4740
062	Extraction of natural gas		129	...	...	...	b/	...	...	...	...	b/	...	...	...	...
071	Mining of iron ores		117	...	...	...	c/	67000	67000	67000	67000	c/	8095	9702	10583	10078
072	Mining of non-ferrous metal ores		993	...	...	...	c/	...	...	...	...	c/	...	...	...	...
081	Quarrying of stone, sand and clay		1286	...	...	...	d/	13000	13000	12000	12000	d/	1018	985	989	1000
089	Mining and quarrying n.e.c.		661	...	...	...	d/	...	...	...	...	d/	...	...	...	...
091	Support activities for petroleum		2092	...	...	...	e/	51000	49000	44000	35000	e/	4998	5149	4864	4717
099	Support activities for other mining, quarrying		3373	...	...	...	e/	...	...	...	...	e/	...	...	...	...
B	Mining and quarrying		9571	...	...	...		194000	192000	190000	178000		23643	25832	27069	26425
351	Electric power generation, transmission		484	...	...	...		53000	52000	50000	48000		5879	5893	5915	6067
352	Manufacture of gas		268	...	...	...		3000	3000	3000	3000		222	236	222	225
353	Steam and air conditioning supply		219	...	...	...		...	...	...	...		...	...	...	...
D	Electricity, gas, steam & air conditioning		971	...	...	...		...	...	...	...		...	...	...	...
360	Water collection, treatment and supply		219	...	...	...	f/	27000	27000	29000	29000	f/	2233	2282	2394	2566
370	Sewerage		403	...	...	...	f/	...	...	...	...	f/	...	...	...	...
381	Waste collection		2560	...	...	...	g/	31000	33000	31000	32000	g/	2012	2062	2209	2333
382	Waste treatment and disposal		975	...	...	...	g/	...	...	...	...	g/	...	...	...	...
383	Materials recovery		269	...	...	...	g/	...	...	...	...	g/	...	...	...	...
390	Remediation activities		198	...	...	...		...	...	...	...		...	...	...	...
E	Water supply;sewerage,waste management		4624	...	...	...		...	...	...	...		...	...	...	...

a/ 051 includes 052.
b/ 061 includes 062.
c/ 071 includes 072.
d/ 081 includes 089.
e/ 091 includes 099.
f/ 360 includes 370.
g/ 381 includes 382 and 383.

Australia

ISIC Revision 4		Output (valuation not defined) (millions of Australian Dollars)					Value added (valuation not defined) (millions of Australian Dollars)			
ISIC	Industry	Note	2012	2013	2014	2015	Note	2012	2013	2014	2015
051	Mining of hard coal	a/	58816	49333	49012	45869	a/	25952	17426	18840	15947
052	Mining of lignite	a/	...	...	...	...	a/	...	...	...	...
061	Extraction of crude petroleum	b/	34870	39684	38928	41369	b/	27783	29091	28954	32204
062	Extraction of natural gas	b/	...	...	...	...	b/	...	...	...	...
071	Mining of iron ores	c/	101149	94490	110326	104129	c/	69730	57700	69962	60966
072	Mining of non-ferrous metal ores	c/	...	...	...	...	c/	...	...	...	...
081	Quarrying of stone, sand and clay	d/	6044	5799	6241	6094	d/	2050	1978	2403	2386
089	Mining and quarrying n.e.c.	d/	...	...	...	...	d/	...	...	...	...
091	Support activities for petroleum	e/	16499	15897	17071	15865	e/	7176	7470	7534	6895
099	Support activities for other mining, quarrying	e/	...	...	...	...	e/	...	...	...	...
B	Mining and quarrying		217377	205203	221578	213326		132692	113665	127692	118397
351	Electric power generation, transmission		63996	73988	77660	73347		24200	25899	25710	26298
352	Manufacture of gas		8570	9623	...	10239		1801	1992	2206	2395
353	Steam and air conditioning supply		...	...	...	...		...	...	...	...
D	Electricity, gas, steam & air conditioning		...	...	...	...		...	...	...	...
360	Water collection, treatment and supply	f/	17590	18511	19367	19908	f/	10044	10457	12003	12500
370	Sewerage	f/	...	...	...	...	f/	...	...	...	...
381	Waste collection		11842g/	11581g/	...	12108g/	g/	4095	4076	4322	4529
382	Waste treatment and disposal		...g/	...g/	...	...g/	g/	...	...	...	...
383	Materials recovery		...g/	...g/	...	...g/	g/	...	...	...	...
390	Remediation activities		...	...	...	...		...	...	...	...
E	Water supply;sewerage,waste management		...	...	...	...		...	...	...	...

a/ 051 includes 052.
b/ 061 includes 062.
c/ 071 includes 072.
d/ 081 includes 089.
e/ 091 includes 099.
f/ 360 includes 370.
g/ 381 includes 382 and 383.

Austria

ISIC Revision 4		Number of enterprises					Number of employees					Wages and salaries paid to employees				
			(number)					(number)					(millions of Euros)			
ISIC	Industry	Note	2012	2013	2014	2015	Note	2012	2013	2014	2015	Note	2012	2013	2014	2015
05	Mining of coal and lignite		-	-	-	-		-	-	-	-		-	-	-	-
051	Mining of hard coal		-	-	-	-		-	-	-	-		-	-	-	-
052	Mining of lignite		-	-	-	-		-	-	-	-		-	-	-	-
06	Extraction of crude petroleum & natural gas		2	2	3	2		...	...	...	...		...	...	...	...
061	Extraction of crude petroleum		2	2	2	2		...	...	...	...		...	...	...	...
062	Extraction of natural gas		-	-	1	-		-	-	...	-		-	-	...	-
07	Mining of metal ores		2	2	2	2		...	...	...	...		...	...	...	...
071	Mining of iron ores		2	2	2	2		...	...	...	...		...	...	...	...
072	Mining of non-ferrous metal ores		-	-	-	-		-	-	-	-		-	-	-	-
08	Other mining and quarrying		342	334	328	332		4567	4543	4593	4567		176	177	187	186
081	Quarrying of stone, sand and clay		322	315	309	311		3917	3893	3926	3917		146	147	155	155
089	Mining and quarrying n.e.c.		20	19	19	21		650	650	667	650		29	31	32	31
09	Mining support service activities		9	15	16	12		38	117	142	212		2	5	5	12
091	Support activities for petroleum		4	8	9	7		16	86	103	212a/		1	4	4	12a/
099	Support activities for other mining, quarrying		5	7	7	5		22	31	39	...a/		1	1	1	...a/
B	Mining and quarrying		355	353	349	348		5874	5919	6052	5997		274	280	297	294
35	Electricity, gas, steam & air conditioning		2142	2256	2271	2390		27519	27674	27546	27261		1621	1642	1701	1664
351	Electric power generation, transmission		1165	1259	1259	1358		21931	24352	24158	24150		1355	1482	1528	1510
352	Manufacture of gas		44	41	47	48		2983	1808	1849	1544		150	107	119	97
353	Steam and air conditioning supply		933	956	965	984		2605	1514	1539	1567		115	54	55	56
D	Electricity, gas, steam & air conditioning		2142	2256	2271	2390		27519	27674	27546	27261		1621	1642	1701	1664
36	Water collection, treatment and supply		566	586	585	589		2347	1762	1813	1662		95	71	76	71
37	Sewerage		659	675	671	657		3195	3200	3342	3365		116	119	127	136
38	Waste collection,treatment,disposal activities		773	775	777	767		13269	14371	14456	14629		455	508	524	537
381	Waste collection		317	323	322	315		8901	9834	9769	9900		296	339	347	356
382	Waste treatment and disposal		314	315	316	310		2226	2281	2343	2389		84	89	92	95
383	Materials recovery		142	137	139	142		2142	2256	2344	2340		75	80	85	86
39	Remediation activities		14	14	16	17		100	87	91	91		3	3	3	3
E	Water supply;sewerage,waste management		2012	2050	2049	2030		18911	19420	19702	19747		669	702	730	747

a/ 091 includes 099.

Austria

ISIC Revision 4		Output (valuation not defined)					Value added at factor values				
		Note	(millions of Euros)				Note	(millions of Euros)			
ISIC	Industry		2012	2013	2014	2015		2012	2013	2014	2015
05	Mining of coal and lignite		-	-	-	-		-	-	-	-
051	Mining of hard coal		-	-	-	-		-	-	-	-
052	Mining of lignite		-	-	-	-		-	-	-	-
06	Extraction of crude petroleum & natural gas		...	...	...	...		...	...	...	...
061	Extraction of crude petroleum		...	...	...	...		...	...	...	...
062	Extraction of natural gas		-	-	...	-		-	-	...	-
07	Mining of metal ores		...	...	...	...		...	...	...	...
071	Mining of iron ores		...	...	...	...		...	...	...	...
072	Mining of non-ferrous metal ores		-	-	-	-		-	-	-	-
08	Other mining and quarrying		1034	1083	1090	1055		403	450	448	434
081	Quarrying of stone, sand and clay		817	858	865	839		319	354	365	350
089	Mining and quarrying n.e.c.		217	225	224	216		84	97	84	84
09	Mining support service activities		17	36	37	43		3	10	14	21
091	Support activities for petroleum		13	32	31	43a/		2	8	12	21a/
099	Support activities for other mining, quarrying		4	5	6	...a/		1	2	2	...a/
B	Mining and quarrying		2607	2501	2326	2023		1294	1245	1162	992
35	Electricity, gas, steam & air conditioning		38680	38780	34591	35532		5472	5591	5189	5107
351	Electric power generation, transmission		24773	26214	23294	23026		4663	5125	4766	4644
352	Manufacture of gas		12446	11544	10209	11345		349	167	130	124
353	Steam and air conditioning supply		1461	1022	1088	1162		460	299	293	339
D	Electricity, gas, steam & air conditioning		38680	38780	34591	35532		5472	5591	5189	5107
36	Water collection, treatment and supply		564	470	485	464		363	310	311	299
37	Sewerage		782	797	811	809		443	444	463	454
38	Waste collection,treatment,disposal activities		2972	3300	3404	3184		1034	1115	1195	1177
381	Waste collection		1397	1669	1658	1658		603	712	737	732
382	Waste treatment and disposal		550	554	604	604		248	253	280	275
383	Materials recovery		1025	1077	1142	923		183	151	179	171
39	Remediation activities		37	55	28	26		5	8	7	7
E	Water supply;sewerage,waste management		4355	4622	4728	4484		1845	1877	1977	1938

a/ 091 includes 099.

Azerbaijan

ISIC Revision 4		Number of establishments					Number of employees					Wages and salaries paid to employees				
			(number)					(number)					(thousands of Azerbaijani Manat)			
ISIC	Industry	Note	2012	2013	2014	2015	Note	2012	2013	2014	2015	Note	2012	2013	2014	2015
051	Mining of hard coal		...	...	...	...		...	...	...	...		...	...	...	...
052	Mining of lignite		...	...	...	...		...	...	...	...		...	...	...	...
061	Extraction of crude petroleum		25	26	26	25		20649	21001	20299	19741		404468	404468	488417	583120
062	Extraction of natural gas		3	3	3	6		3050	2959	2889	2712		42034	42034	47534	44081
071	Mining of iron ores		2	2	1	1		362	393	361	361		1511	1511	1447	1779
072	Mining of non-ferrous metal ores		1	1	2	3		525	554	583	595		6085	6086	7306	7666
081	Quarrying of stone, sand and clay		191	198	192	194		2951	5452	3233	2750		8646	8646	15115	9846
089	Mining and quarrying n.e.c.		9	10	11	9		850	860	277	404		3412	3413	1405	2416
091	Support activities for petroleum		58	78	87	87		8434	8450	9092	8383		144671	144671	187522	244773
099	Support activities for other mining, quarrying		...	...	...	...		...	...	...	...		...	...	...	...
B	Mining and quarrying		289	318	322	325		36821	39669	36734	34946		610829	610829	748746	893681
351	Electric power generation, transmission		105	105	106	106		16810	16985	16532	15836		16810	87869	93579	95883
352	Manufacture of gas		74	73	74	74		10453	10631	9862	8690		10453	62033	67276	56835
353	Steam and air conditioning supply		36	42	43	43		2317	2896	3125	2454		78297	7132	11516	8837
D	Electricity, gas, steam & air conditioning		215	220	223	223		29580	30512	29519	26980		29580	157034	172372	161555
360	Water collection, treatment and supply		68	81	83	83		9450	11323	10796	10870		37459	37459	54772	53735
370	Sewerage		72	16	18	18		4448	774	865	887		4448	9199	1699	1736
381	Waste collection		15	61	59	59		2670	5488	5662	5726		2670	8455	13793	14187
382	Waste treatment and disposal		9	10	9	9		593	560	556	558		593	984	1053	1231
383	Materials recovery		7	8	7	7		172	163	207	191		172	762	952	1041
390	Remediation activities		44	49	51	51		6450	7165	7698	7259		6450	21434	29764	28410
E	Water supply;sewerage,waste management		215	225	227	227		23783	25473	25784	25491		78292	78293	102033	100340

Azerbaijan

ISIC Revision 4		Output at factor values					Value added at factor values				
		Note	(thousands of Azerbaijani Manat)				Note	(thousands of Azerbaijani Manat)			
ISIC	Industry		2012	2013	2014	2015		2012	2013	2014	2015
051	Mining of hard coal		...	...	...	...		...	...	...	...
052	Mining of lignite		...	...	...	...		...	...	...	...
061	Extraction of crude petroleum		24668206	23524973	20851236	14588351		23138473	22299361	19475328	13582819
062	Extraction of natural gas		78758	132849	125460	134970		68379	120095	60221	113240
071	Mining of iron ores		4361	3515	1844	4395		1295	1651	1210	1560
072	Mining of non-ferrous metal ores		63770	47726	53142	83024		21448	15574	29440	27644
081	Quarrying of stone, sand and clay		48610	46844	39416	44506		15388	15221	16510	14418
089	Mining and quarrying n.e.c.		17779	18430	9605	9278		5591	7168	5653	4569
091	Support activities for petroleum		725735	880997	900227	1497791		319623	331105	584860	564020
099	Support activities for other mining, quarrying		...	...	...	...		...	...	...	...
B	Mining and quarrying		25607219	24655334	21980930	16362315		23570197	22790175	20173221	14308270
351	Electric power generation, transmission		722111	726150	865746	928838		432311	443265	464412	374208
352	Manufacture of gas		983516	1042155	951373	949010		640820	660441	549805	358362
353	Steam and air conditioning supply		18716	5554	7284	9374		9116	3824	452	2113
D	Electricity, gas, steam & air conditioning		1724343	1773859	1824403	1887222		1082247	1107530	1014669	734683
360	Water collection, treatment and supply		112662	129346	129296	140453		25068	61164	42048	64556
370	Sewerage		14521	22752	4269	1458		6358	4236	1622	529
381	Waste collection		2814	...	22272	26144		1619	2377	10811	15087
382	Waste treatment and disposal		18043	7910	7273	5268		4156	1683	3073	1825
383	Materials recovery		3185	905	124	158		2428	534	114	128
390	Remediation activities		50484	59833	70129	65969		26643	21083	13297	11330
E	Water supply;sewerage,waste management		201708	225129	233363	239450		66272	91077	70964	93455

Bangladesh

ISIC Revision 4		Establishments					Number of employees					Wages and salaries				
			(number)					(number)					(millions of Bangladeshi Taka)			
ISIC	Industry	Note	2012	2013	2014	2015	Note	2012	2013	2014	2015	Note	2012	2013	2014	2015
05	Mining of coal and lignite		...	...	...	...		...	...	...	1000		...	...	...	...
06	Extraction of crude petroleum & natural gas		...	...	...	...		...	...	...	1000		...	...	...	...
07	Mining of metal ores		...	...	...	...		...	...	...	2000		...	...	...	...
08	Other mining and quarrying		...	...	...	...		...	...	...	84000		...	...	...	...
09	Mining support service activities		...	...	...	...		...	...	...	3000		...	...	...	...
B	Mining and quarrying		...	...	...	...	a/	...	...	...	91000		...	...	...	...
35	Electricity, gas, steam & air conditioning		...	...	...	...		...	...	...	111000		...	...	...	...
D	Electricity, gas, steam & air conditioning		...	...	...	...	a/	...	...	...	111000		...	...	...	...
36	Water collection, treatment and supply		...	...	...	...		...	...	...	41000		...	...	...	...
37	Sewerage		...	...	...	...		...	...	...	6000		...	...	...	...
38	Waste collection,treatment,disposal activities		...	...	...	...		...	...	...	6000		...	...	...	...
39	Remediation activities		...	...	...	...		...	...	...	1000		...	...	...	...
E	Water supply;sewerage,waste management		...	...	...	...	a/	...	...	...	54000		...	...	...	...

a/ Sum of available data.

Bangladesh

ISIC Revision 4		Output at basic prices					Value added at basic prices				
			(millions of Bangladeshi Taka)					(millions of Bangladeshi Taka)			
ISIC	Industry	Note	2012	2013	2014	2015	Note	2012	2013	2014	2015
05	Mining of coal and lignite		...	8767	6897	11420		...	6823	5367	8887
051	Mining of hard coal		...	8767	6897	11420		...	6823	5367	8887
052	Mining of lignite		...	...	...	...		...	...	...	...
06	Extraction of crude petroleum & natural gas		...	94381	106461	120634		...	83026	93609	109103
061	Extraction of crude petroleum		...	...	...	...		...	...	...	...
062	Extraction of natural gas		...	...	...	...		...	...	...	...
07	Mining of metal ores		...	...	...	...		...	...	...	...
071	Mining of iron ores		...	...	...	...		...	...	...	...
072	Mining of non-ferrous metal ores		...	...	...	...		...	...	...	...
08	Other mining and quarrying		...	138782	160646	191693		...	122420	141513	169831
081	Quarrying of stone, sand and clay		...	121904	143570	174281		...	107230	126144	154160
089	Mining and quarrying n.e.c.		...	16878	17076	17412		...	15190	15369	15671
09	Mining support service activities		...	...	...	...		...	...	...	...
091	Support activities for petroleum		...	...	...	...		...	...	...	...
099	Support activities for other mining, quarrying		...	...	...	...		...	...	...	...
B	Mining and quarrying	a/	...	241930	274004	323747	a/	...	212269	240489	287821
35	Electricity, gas, steam & air conditioning		...	315371	321373	380855		...	150524	192309	231777
351	Electric power generation, transmission		...	264172	271354	322188		...	138833	151202	185159
352	Manufacture of gas		...	51198	50019	58666		...	11691	41107	46618
353	Steam and air conditioning supply		...	...	...	...		...	...	...	...
D	Electricity, gas, steam & air conditioning	a/	...	315371	321373	380855	a/	...	150524	192309	231777
36	Water collection, treatment and supply		...	14347	16083	18000		...	9631	11051	12033
37	Sewerage		...	...	...	...		...	...	...	...
38	Waste collection,treatment,disposal activities		...	...	...	...		...	...	...	...
381	Waste collection		...	...	...	...		...	...	...	...
382	Waste treatment and disposal		...	...	...	...		...	...	...	...
383	Materials recovery		...	...	...	...		...	...	...	...
39	Remediation activities		...	...	...	...		...	...	...	...
E	Water supply;sewerage,waste management	a/	...	14347	16083	18000	a/	...	9631	11051	12033

a/ Sum of available data.

Belarus

ISIC Revision 4		Number of establishments					Number of employees					Wages and salaries paid to employees				
			(number)					(thousands)					(billions of Belarusian Roubles)			
ISIC	Industry	Note	2012	2013	2014	2015	Note	2012	2013	2014	2015	Note	2012	2013	2014	2015
B	Mining and quarrying		...	37	39	33		...	11.9	11.4	10.9		...	1198.0	1304.6	1312.4
D	Electricity, gas, steam & air conditioning		...	211	180	182		...	98.1	98.2	97.8		...	6458.2	7915.6	8885.8
E	Water supply;sewerage,waste management		...	194	200	243		...	38.8	38.8	39.0		...	2251.9	2653.0	2862.3

Belarus

ISIC Revision 4		Output at basic prices					Value added at basic prices				
			(billions of Belarusian Roubles)					(billions of Belarusian Roubles)			
ISIC	Industry	Note	2012	2013	2014	2015	Note	2012	2013	2014	2015
B	Mining and quarrying		...	8790.6	8989.6	9023.3		...	6115.1	5469.6	5237.8
D	Electricity, gas, steam & air conditioning		...	46437.7	55870.0	74060.3		...	15241.8	20504.8	26146.9
E	Water supply;sewerage,waste management		...	10334.8	11852.5	12729.5		...	4432.2	5807.9	6450.1

Belgium

ISIC Revision 4		Number of enterprises					Number of employees					Wages and salaries paid to employees				
		Note	(number)				Note	(number)				Note	(millions of Euros)			
ISIC	Industry		2012	2013	2014	2015		2012	2013	2014	2015		2012	2013	2014	2015
05	Mining of coal and lignite		...	-	-	-		...	-	-	-		...	-	-	-
051	Mining of hard coal		-	-	-	-		-	-	-	-		-	-	-	-
052	Mining of lignite		...	-	-	-		...	-	-	-		...	-	-	-
06	Extraction of crude petroleum & natural gas		-	-	-	-		-	-	-	-		-	-	-	-
061	Extraction of crude petroleum		-	-	-	-		-	-	-	-		-	-	-	-
062	Extraction of natural gas		-	-	-	-		-	-	-	-		-	-	-	-
07	Mining of metal ores		-	-	-	-		-	-	-	-		-	-	-	-
071	Mining of iron ores		-	-	-	-		-	-	-	-		-	-	-	-
072	Mining of non-ferrous metal ores		-	-	-	-		-	-	-	-		-	-	-	-
08	Other mining and quarrying		...	177	175	...		...	2525	2666	...		...	98	115	...
081	Quarrying of stone, sand and clay		166	169	171	177		2473	2350	2501	2301		100	92	109	95
089	Mining and quarrying n.e.c.		...	8	4	...		...	175	165	...		...	7	6	...
09	Mining support service activities		31	37	17	...		16	24	17	...		1	2	1	...
091	Support activities for petroleum		9	37a/	17a/	...		1	24a/	17a/	...		-	2a/	1a/	...
099	Support activities for other mining, quarrying		22	...a/	...a/	15		15	...a/	...a/	5		1	...a/	...a/	-
B	Mining and quarrying		203	214	192	213		2598	2549	2683	2506		105	100	116	103
35	Electricity, gas, steam & air conditioning		533	618	693	851		18007	19510	20166	19550		1445	1549	1498	1505
351	Electric power generation, transmission		501	577	655	826		16740	18076	17706	19207		1343	1437	1328	1474
352	Manufacture of gas		16	18	17	12		1251	1418	2436	331		102	112	170	31
353	Steam and air conditioning supply		16	23	21	13		16	16	24	12		-	1	1	-
D	Electricity, gas, steam & air conditioning		533	618	693	851		18007	19510	20166	19550		1445	1549	1498	1505
36	Water collection, treatment and supply		40	31	35	46		8184	7541	7444	7342		422	398	400	395
37	Sewerage		266	177	150	217		1863	2188	1848	1917		83	102	94	99
38	Waste collection,treatment,disposal activities		947	937	980	902		14362	15217	11034	11727		539	587	412	436
381	Waste collection		174	186	155	181		4157	4198	3160	4348		143	149	108	148
382	Waste treatment and disposal		318	276	290	278		6731	7353	4117	3815		284	313	174	172
383	Materials recovery		455	475	535	443		3474	3666	3757	3564		112	124	131	117
39	Remediation activities		76	95	88	111		1322	1298	1116	1439		59	58	47	58
E	Water supply;sewerage,waste management		1329	1240	1253	1276		25731	26244	21442	22425		1103	1146	953	989

a/ 091 includes 099.

Belgium

ISIC Revision 4		Output (valuation not defined)					Value added at factor values				
		Note	(millions of Euros)				Note	(millions of Euros)			
ISIC	Industry		2012	2013	2014	2015		2012	2013	2014	2015
05	Mining of coal and lignite		...	-	-	-		...	-	-	-
051	Mining of hard coal		-	-	-	-		-	-	-	-
052	Mining of lignite		...	-	-	-		...	-	-	-
06	Extraction of crude petroleum & natural gas		-	-	-	-		-	-	-	-
061	Extraction of crude petroleum		-	-	-	-		-	-	-	-
062	Extraction of natural gas		-	-	-	-		-	-	-	-
07	Mining of metal ores		-	-	-	-		-	-	-	-
071	Mining of iron ores		-	-	-	-		-	-	-	-
072	Mining of non-ferrous metal ores		-	-	-	-		-	-	-	-
08	Other mining and quarrying		...	823	879	...		...	257	290	...
081	Quarrying of stone, sand and clay		779	716	796	609		256	242	269	211
089	Mining and quarrying n.e.c.		...	107	83	...		...	16	21	...
09	Mining support service activities		18	24	24	...		6	2	2	...
091	Support activities for petroleum		11	24a/	24a/	...		3	2a/	2a/	...
099	Support activities for other mining, quarrying		6	...a/	...a/	1		3	...a/	...a/	-
B	Mining and quarrying		857	847	903	811		271	259	292	236
35	Electricity, gas, steam & air conditioning		41252	42099	39171	35431		5677	7455	5874	5279
351	Electric power generation, transmission		35306	34830	34415	32765		5153	5315	5079	5090
352	Manufacture of gas		5889	7214	4708	2629		...	...	...	...
353	Steam and air conditioning supply		57	55	48	37		5	5	5	2
D	Electricity, gas, steam & air conditioning		41252	42099	39171	35431		5677	7455	5874	5279
36	Water collection, treatment and supply		2761	2661	2535	2834		1158	1088	1083	1143
37	Sewerage		730	1167	1092	1117		410	580	541	580
38	Waste collection,treatment,disposal activities		5749	5804	5154	4937		1373	1414	1016	1082
381	Waste collection		743	884	724	1131		262	340	237	317
382	Waste treatment and disposal		2178	2179	1625	1490		779	751	456	466
383	Materials recovery		2828	2741	2806	2316		331	323	322	300
39	Remediation activities		425	393	359	356		60	61	103	123
E	Water supply;sewerage,waste management		9665	10025	9139	9244		3000	3143	2742	2929

a/ 091 includes 099.

Bermuda

ISIC Revision 3		Number of establishments					Number of persons engaged					Wages and salaries				
			(number)					(number)					(Bermuda Dollars)			
ISIC	Industry	Note	2012	2013	2014	2015	Note	2012	2013	2014	2015	Note	2012	2013	2014	2015
101	Mining and agglomeration of hard coal		...	...	...	...		...	...	...	...		...	...	...	...
102	Mining and agglomeration of lignite		...	...	...	...		...	...	...	...		...	...	...	...
103	Extraction and agglomeration of peat		...	...	...	...		...	...	...	...		...	...	...	...
111	Extraction of crude oil and natural gas		...	...	...	...		...	...	...	...		...	...	...	...
112	Oil, gas service activities		...	...	...	...		...	...	...	...		...	...	...	...
120	Mining of uranium and thorium ores		...	...	...	...		...	...	...	...		...	...	...	...
131	Iron ores		...	...	...	...		...	...	...	...		...	...	...	...
132	Non-ferrous metal ores,excl.uranium,thorium		...	...	...	...		...	...	...	...		...	...	...	...
141	Quarrying of stone, sand and clay		1	...	1	...		1	...	1	...		...	...	...	...
142	Mining and quarrying n.e.c.		...	...	...	...		...	...	...	...		...	...	...	...
C	Total mining and quarrying		1a/	...	1a/	...		1a/	...	1a/	...		...	...	...	...
401	Electricity production, collection, distribution		1	...	2	2		315	...	259	258		...	...	...	...
402	Gas production and distribution		...	...	1	1		...	...	1	2		...	...	...	...
403	Steam and hot water supply		...	...	...	...		...	...	...	...		...	...	...	...
410	Water collection, purification, distribution		21	...	20	20		71	...	65	61		...	...	...	...
E	Total electricity, gas and water supply	a/	22	...	23	23	a/	386	...	325	321		...	...	...	...

a/ Sum of available data.

Botswana

ISIC Revision 3		Establishments					Employees					Wages and salaries paid to employees				
			(number)					(number)					(millions of Botswana Pula)			
ISIC	Industry	Note	2012	2013	2014	2015	Note	2012	2013	2014	2015	Note	2012	2013	2014	2015
101	Mining and agglomeration of hard coal		...	...	...	...		...	...	...	...	a/	87	101	122	157
102	Mining and agglomeration of lignite		...	...	...	...		...	...	...	...	a/	...	...	...	...
103	Extraction and agglomeration of peat		...	...	...	...		...	...	...	...	a/	...	...	...	...
111	Extraction of crude oil and natural gas		...	...	...	...		...	...	...	...		...	...	...	...
112	Oil, gas service activities		...	...	...	...		...	...	...	...		...	...	...	...
120	Mining of uranium and thorium ores		...	...	...	...		...	...	...	...		...	...	...	...
131	Iron ores		...	...	...	...		...	...	...	...		...	...	...	...
132	Non-ferrous metal ores,excl.uranium,thorium		...	...	...	...		...	...	...	...		883	914	1004	1008
141	Quarrying of stone, sand and clay		...	...	...	...		...	...	...	...		...	...	...	...
142	Mining and quarrying n.e.c.		...	...	...	...		...	...	...	...		1734	1919	2406	2317
C	Total mining and quarrying		...	...	...	...		...	...	...	...		...	...	...	...
401	Electricity production, collection, distribution		...	...	...	...		...	...	...	...		419	406	459	499
402	Gas production and distribution		...	...	...	...		...	...	...	...		...	...	...	...
403	Steam and hot water supply		...	...	...	...		...	...	...	...		...	...	...	...
410	Water collection, purification, distribution		...	...	...	...		...	...	...	...		653	666	679	662
E	Total electricity, gas and water supply		...	...	...	...		...	...	...	...	b/	1073	1072	1139	1161

a/ 101 includes 102 and 103.
b/ E includes electricity and water only.

Botswana

ISIC Revision 3		Output at basic prices					Value added at basic prices				
			(millions of Botswana Pula)					(millions of Botswana Pula)			
ISIC	Industry	Note	2012	2013	2014	2015	Note	2012	2013	2014	2015
101	Mining and agglomeration of hard coal	a/	321	615	424	476	a/	246	496	182	204
102	Mining and agglomeration of lignite	a/	...	...	...	...	a/	...	...	...	...
103	Extraction and agglomeration of peat	a/	...	...	...	...	a/	...	...	...	...
111	Extraction of crude oil and natural gas		...	...	...	...		...	...	...	...
112	Oil, gas service activities		...	...	...	...		...	...	...	...
120	Mining of uranium and thorium ores		...	...	...	...		...	...	...	...
131	Iron ores		...	...	...	...		...	...	...	...
132	Non-ferrous metal ores,excl.uranium,thorium		4342	5643	5421	4807		1492	1284	1160	1488
141	Quarrying of stone, sand and clay		...	...	...	...		...	...	...	...
142	Mining and quarrying n.e.c.		23104	29600	38126	28649		17111	21986	30506	23556
C	Total mining and quarrying		27767	35857	43971	33931		19286	24243	32403	25820
401	Electricity production, collection, distribution		2080	2306	2606	2950		...	...	...	...
402	Gas production and distribution		...	...	...	...		...	...	...	...
403	Steam and hot water supply		...	...	...	...		...	...	...	...
410	Water collection, purification, distribution		1036	1100	1095	1423		644	660	611	742
E	Total electricity, gas and water supply	b/	3115	3406	3701	4372	b/	-621	-123	-582	-219

a/ 101 includes 102 and 103.
b/ E includes electricity and water only.

Brazil

ISIC Revision 4		Number of enterprises					Number of employees					Wages and salaries paid to employees				
			(number)					(number)					(millions of Brazilian Reais)			
ISIC	Industry	Note	2012	2013	2014	2015	Note	2012	2013	2014	2015	Note	2012	2013	2014	2015
051	Mining of hard coal	a/	19	19	19	15	a/	5464	5739	5725	4729	a/	198.1	217.5	273.7	279.7
052	Mining of lignite	a/	...	...	...	...	a/	...	...	...	...	a/	...	...	...	...
061	Extraction of crude petroleum	b/	33	29	32	33	b/	2853	3170	3314	3010	b/	734.5	828.1	786.8	817.9
062	Extraction of natural gas	b/	...	...	...	...	b/	...	...	...	...	b/	...	...	...	...
071	Mining of iron ores		52	53	53	51		69343	69227	67658	65771		4106.1	4452.4	4496.7	2933.6
072	Mining of non-ferrous metal ores		111	112	118	99		15480	13779	14220	13673		704.5	723.1	826.8	844.2
081	Quarrying of stone, sand and clay		2699	3162	3051	3106		60796	68402	66826	64067		1311.3	1601.9	1710.1	1792.0
089	Mining and quarrying n.e.c.		348	369	339	327		15901	15746	15156	14758		408.0	404.7	435.0	444.1
091	Support activities for petroleum		93	95	103	99		36829	34441	38652	33296		3407.5	4221.3	4223.9	4409.4
099	Support activities for other mining, quarrying		23	24	20	30		2287	2415	2075	3022		57.9	67.8	72.1	88.7
B	Mining and quarrying		3378	3863	3735	3760		208953	212919	213626	202326		10927.9	12516.8	12825.2	11609.6
351	Electric power generation, transmission		...	...	...	...		...	...	...	...		...	...	...	...
352	Manufacture of gas		...	...	...	...		...	...	...	...		...	...	...	...
353	Steam and air conditioning supply		...	...	...	...		...	...	...	...		...	...	...	...
D	Electricity, gas, steam & air conditioning		...	...	...	...		...	...	...	...		...	...	...	...
360	Water collection, treatment and supply		...	...	...	...		...	...	...	...		...	...	...	...
370	Sewerage		...	...	...	...		...	...	...	...		...	...	...	...
381	Waste collection		...	...	...	...		...	...	...	...		...	...	...	...
382	Waste treatment and disposal		...	...	...	...		...	...	...	...		...	...	...	...
383	Materials recovery		...	...	...	...		...	...	...	...		...	...	...	...
390	Remediation activities		...	...	...	...		...	...	...	...		...	...	...	...
E	Water supply;sewerage,waste management		...	...	...	...		...	...	...	...		...	...	...	...

a/ 051 includes 052.
b/ 061 includes 062.

Brazil

ISIC Revision 4		Output at factor values					Value added at factor values				
			(millions of Brazilian Reais)					(millions of Brazilian Reais)			
ISIC	Industry	Note	2012	2013	2014	2015	Note	2012	2013	2014	2015
051	Mining of hard coal	a/	1113.9	1323.8	1603.1	1385.2	a/	383.2	485.1	534.9	512.8
052	Mining of lignite	a/	...	...	...	...	a/	...	...	...	...
061	Extraction of crude petroleum	b/	13602.9	17163.8	25331.8	21189.5	b/	4781.6	6013.5	9912.4	5780.7
062	Extraction of natural gas	b/	...	...	...	...	b/	...	...	...	...
071	Mining of iron ores		82399.2	87981.3	70547.3	57388.2		61775.5	67028.7	44916.6	34370.9
072	Mining of non-ferrous metal ores		6495.4	6732.2	7399.4	10187.6		2548.8	2431.7	2622.9	4802.1
081	Quarrying of stone, sand and clay		9555.0	11334.4	11529.7	10914.7		4793.0	5664.2	5716.9	5239.0
089	Mining and quarrying n.e.c.		3151.3	2855.3	2980.7	3023.6		1307.4	1216.9	1230.3	1173.4
091	Support activities for petroleum		11923.4	13262.7	14412.5	16495.8		6210.0	6415.3	6871.7	7983.8
099	Support activities for other mining, quarrying		286.5	310.2	310.1	388.9		179.9	169.0	174.0	225.3
B	Mining and quarrying		128527.5	140963.6	134114.5	120973.6		81979.4	89424.4	71979.6	60088.0
351	Electric power generation, transmission		...	...	...	...		...	...	...	...
352	Manufacture of gas		...	...	...	...		...	...	...	...
353	Steam and air conditioning supply		...	...	...	...		...	...	...	...
D	Electricity, gas, steam & air conditioning		...	...	...	...		...	...	...	...
360	Water collection, treatment and supply		...	...	...	...		...	...	...	...
370	Sewerage		...	...	...	...		...	...	...	...
381	Waste collection		...	...	...	...		...	...	...	...
382	Waste treatment and disposal		...	...	...	...		...	...	...	...
383	Materials recovery		...	...	...	...		...	...	...	...
390	Remediation activities		...	...	...	...		...	...	...	...
E	Water supply;sewerage,waste management		...	...	...	...		...	...	...	...

a/ 051 includes 052.
b/ 061 includes 062.

Bulgaria

ISIC Revision 4		Number of establishments					Number of employees					Wages and salaries paid to employees				
			(number)					(number)					(thousands of Bulgarian Leva)			
ISIC	Industry	Note	2012	2013	2014	2015	Note	2012	2013	2014	2015	Note	2012	2013	2014	2015
051	Mining of hard coal		3	4	4	4		153	128	107	74		1170	1083	898	658
052	Mining of lignite		21	20	20	18		12766	12375	12007	11912		197008	187304	191225	197054
061	Extraction of crude petroleum	a/	7	6	7	4		...	...	...	...		...	7744a/	8029a/	...
062	Extraction of natural gas	a/	...	...	...	...		46	57	46	34		1510	...a/	...a/	1304
071	Mining of iron ores		...	...	...	...		...	...	...	...		...	...	...	...
072	Mining of non-ferrous metal ores		24	26	23	23		6718	6721	6846	6693		120705	124944	135752	132716
081	Quarrying of stone, sand and clay		259	266	246	259		3736	3700	3354	4549		31772	31742	30663	49377
089	Mining and quarrying n.e.c.		32	30	25	28		478	464	351	371		4739	4910	4650	4463
091	Support activities for petroleum		12	11	12	14		25	35	36	21		902	1366	1017	655
099	Support activities for other mining, quarrying		24	23	22	24		159	182	332	311		1906	2181	3967	4671
B	Mining and quarrying		382	386	359	374		24641	24268	23709	24617		364644	361274	376201	397060
351	Electric power generation, transmission		1631	1714	1670	1683		27464	27148	26991	26137		531945	524369	533412	541621
352	Manufacture of gas		39	38	42	33		755	713	687	753		12695	11500	11695	12950
353	Steam and air conditioning supply		33	32	31	29		3593	3476	3517	3358		41961	42345	44329	39115
D	Electricity, gas, steam & air conditioning		1703	1784	1743	1745		31812	31337	31195	30248		586601	578214	589436	593686
360	Water collection, treatment and supply		71	70	72	65		17183	17137	17004	16925		145488	151429	157794	161194
370	Sewerage		67	64	65	65		555	560	549	580		3709	4246	4235	4630
381	Waste collection		328	320	322	349		11932	12031	11270	11498		80773	83243	80732	92696
382	Waste treatment and disposal		80	75	78	99		2168	2535	2605	3402		23456	34770	38019	42557
383	Materials recovery		149	161	150	146		2516	2241	1788	1573		22825	19235	16624	14545
390	Remediation activities		53	61	58	64		711	718	814	962		6230	7143	7808	9126
E	Water supply;sewerage,waste management		748	751	745	788		35065	35222	34030	34940		282481	300066	305212	324748

a/ 061 includes 062.

Bulgaria

ISIC Revision 4			Output at producers' prices					Value added at factor values			
			(millions of Bulgarian Leva)					(millions of Bulgarian Leva)			
ISIC	Industry	Note	2012	2013	2014	2015	Note	2012	2013	2014	2015
051	Mining of hard coal		6	6	6	6		2	328a/	367a/	1
052	Mining of lignite		709	587	645	754		358	...a/	...a/	452
061	Extraction of crude petroleum	b/	233	174	132	67	b/	183	110	80	31
062	Extraction of natural gas	b/	...	...	...	...	b/	...	...	...	...
071	Mining of iron ores		...	...	...	...		...	...	...	...
072	Mining of non-ferrous metal ores		1743	1505	1332	1237		1021	767	596	495
081	Quarrying of stone, sand and clay		232	230	251	445		83	78	96	165
089	Mining and quarrying n.e.c.		28	36	38	42		8	9	12	10
091	Support activities for petroleum		4	3	2	1		1	2	1c/	1c/
099	Support activities for other mining, quarrying		15	16	26	26		4	5	...c/	...c/
B	Mining and quarrying		2970	2557	2433	2579		1660	1299	1152	1155
351	Electric power generation, transmission		8475	7808	6206	6988		2772	2979	2305	2912
352	Manufacture of gas		...	167	104	136		-37	112	61	106
353	Steam and air conditioning supply		1167	1068	965	775		225	232	175	118
D	Electricity, gas, steam & air conditioning		9646	9043	7275	7899		2960	3323	2541	3136
360	Water collection, treatment and supply		639	643	618	643		336	335	335	354
370	Sewerage		12	13	17	16		5	5	7	7
381	Waste collection		351	333	337	348		159	161	162	176
382	Waste treatment and disposal		86	92	95	139		56	80	89	108
383	Materials recovery		226	184	176	156		95	77	64	72
390	Remediation activities		60	93	112	118		15	17	21	24
E	Water supply;sewerage,waste management		1374	1358	1355	1420		666	675	678	741

a/ 051 includes 052.
b/ 061 includes 062.
c/ 091 includes 099.

Burundi

ISIC Revision 3		Number of enterprises					Number of employees					Wages and salaries paid to employees				
			(number)					(number)					(millions of Burundi Francs)			
ISIC	Industry	Note	2011	2012	2013	2014	Note	2011	2012	2013	2014	Note	2011	2012	2013	2014
101	Mining and agglomeration of hard coal		...	...	...	...		...	...	...	...		...	...	...	...
102	Mining and agglomeration of lignite		...	...	...	...		...	...	...	...		...	...	...	...
103	Extraction and agglomeration of peat		...	...	...	...		...	...	...	...		...	...	...	...
111	Extraction of crude oil and natural gas		...	...	...	...		...	...	...	...		...	...	...	...
112	Oil, gas service activities		...	...	...	...		...	...	...	...		...	...	...	...
120	Mining of uranium and thorium ores		...	...	...	...		...	...	...	...		...	...	...	...
131	Iron ores		...	...	...	...		...	...	...	...		...	...	...	...
132	Non-ferrous metal ores,excl.uranium,thorium		...	...	...	...		...	...	...	...		...	...	...	...
141	Quarrying of stone, sand and clay		...	...	...	...		...	...	...	...		...	...	...	...
142	Mining and quarrying n.e.c.		...	...	...	...		...	...	...	...		...	...	...	...
C	Total mining and quarrying		...	...	...	...		...	...	...	...		...	...	...	...
401	Electricity production, collection, distribution		1	1	...	...		1015	1038	1076	...		6171	6310	...	...
402	Gas production and distribution		...	...	...	...		...	...	...	...		...	...	...	...
403	Steam and hot water supply		...	...	...	...		...	...	...	...		...	...	...	...
410	Water collection, purification, distribution		1	1	...	...		435	445	461	...		2270	3559	...	...
E	Total electricity, gas and water supply	a/	2	2	...	...	a/	1450	1483	1537	...	a/	8441	9869	...	...

a/ Sum of available data.

Burundi

ISIC Revision 3		Output at basic prices					Value added at basic prices				
			(millions of Burundi Francs)					(millions of Burundi Francs)			
ISIC	Industry	Note	2011	2012	2013	2014	Note	2011	2012	2013	2014
101	Mining and agglomeration of hard coal		...	...	...	...		...	...	...	...
102	Mining and agglomeration of lignite		...	...	...	...		...	...	...	...
103	Extraction and agglomeration of peat		...	...	...	...		...	...	...	...
111	Extraction of crude oil and natural gas		...	...	...	...		...	...	...	...
112	Oil, gas service activities		...	...	...	...		...	...	...	...
120	Mining of uranium and thorium ores		...	...	...	...		...	...	...	...
131	Iron ores		...	...	...	...		...	...	...	...
132	Non-ferrous metal ores,excl.uranium,thorium		...	...	...	...		...	...	...	...
141	Quarrying of stone, sand and clay		...	...	...	...		...	...	...	...
142	Mining and quarrying n.e.c.		...	...	...	...		...	...	...	...
C	Total mining and quarrying		...	...	...	...		...	...	...	...
401	Electricity production, collection, distribution		17940	26043	28501	...		8565	15057	16456	...
402	Gas production and distribution		...	...	...	...		...	...	...	...
403	Steam and hot water supply		...	...	...	...		...	...	...	...
410	Water collection, purification, distribution		11960	17362	19001	...		6133	10038	10972	...
E	Total electricity, gas and water supply	a/	29900	43405	47502	...	a/	14698	25095	27428	...

a/ Sum of available data.

China

ISIC Revision 4		Number of enterprises					Number of employees					Wages and salaries paid to employees			
		Note	(number)				Note	(thousands)				Note	(millions of Chinese Yuan)		
ISIC	Industry		2012	2013	2014	2015		2012	2013	2014	2015		2012	2013	2014	2015
051	Mining of hard coal		7618	7678	6613	5706		...	5139	4698	4191		...	287814	265210	224502
052	Mining of lignite		218	216	198	178		...	172	167	159		...	9709	11365	11101
061	Extraction of crude petroleum		103	108	105	93		...	734	716	702		...	56666	59628	58550
062	Extraction of natural gas		31	36	37	48		...	17	18	21		...	1451	1638	1849
071	Mining of iron ores		3256	3359	3096	2418		...	664	620	488		...	28277	27224	21975
072	Mining of non-ferrous metal ores		2318	2363	2218	2064		...	604	569	503		...	25131	24065	22307
081	Quarrying of stone, sand and clay		2413	2679	2741	2781		...	331	328	321		...	11424	12506	13222
089	Mining and quarrying n.e.c.		1018	1092	1075	1022		...	217	214	196		...	8375	8924	8638
091	Support activities for petroleum		138	160	150	154		...	319	312	298		...	22101	24225	22658
099	Support activities for other mining, quarrying		21	20	16	15		...	2	2	2		...	66	102	87
B	Mining and quarrying		17134	17711	16249	14479		...	8198	7643	6881		...	451012	434887	384888
351	Electric power generation, transmission		4739	5065	5442	5847		...	2651	2616	2583		...	553684	590068	647436
352	Manufacture of gas		990	1132	1308	1416		...	242	265	275		...	56058	60617	64800
353	Steam and air conditioning supply		875	945	1029	1120		...	233	257	262		...	41162	45879	50497
D	Electricity, gas, steam & air conditioning		6604	7142	7779	8383		...	3126	3139	3119		...	650904	696564	762733
360	Water collection, treatment and supply		1018	1093	1158	1229		...	363	370	382		...	108031	136026	145366
370	Sewerage		241	283	337	392		...	33	38	45		...	53542	58415	62741
381	Waste collection		...	...	...	...		...	...	...	...		...	...	...	...
382	Waste treatment and disposal		...	...	...	...		...	...	...	...		...	...	...	...
383	Materials recovery		1293	1453	1563	1599		...	200	198	193		...	135743	142102	158637
390	Remediation activities		...	...	...	...		...	...	...	...		...	...	...	...
E	Water supply;sewerage,waste management		2552	2829	3058	3220		...	597	606	620		...	297316	336543	366744

China

ISIC Revision 4		Output (valuation not defined)					Value added (valuation not defined)			
		Note	(billions of Chinese Yuan)				Note	(billions of Chinese Yuan)		
ISIC	Industry		2012	2013	2014	2015		2012	2013	2014	2015
051	Mining of hard coal		3116.4	2979.9	2732.9	2183.1		...	...	...	...
052	Mining of lignite		282.2	308.3	291.5	187.2		...	...	...	...
061	Extraction of crude petroleum		1071.3	1051.1	1024.6	620.2		...	...	...	...
062	Extraction of natural gas		95.2	107.9	117.9	170.6		...	...	...	...
071	Mining of iron ores		839.1	944.1	890.6	682.5		...	...	...	...
072	Mining of non-ferrous metal ores		602.0	660.7	673.2	661.8		...	...	...	...
081	Quarrying of stone, sand and clay		282.7	330.8	363.4	379.5		...	...	...	...
089	Mining and quarrying n.e.c.		147.0	169.8	175.6	171.2		...	...	...	...
091	Support activities for petroleum		183.6	202.4	208.2	172.6		...	...	...	...
099	Support activities for other mining, quarrying		3.3	1.5	1.8	1.4		...	...	...	...
B	Mining and quarrying		6622.8	6756.4	6479.6	5230.0		2509.3	2546.8	2341.7	1910.5
351	Electric power generation, transmission		5151.6	5471.6	5547.6	5493.1		...	...	...	...
352	Manufacture of gas		335.9	405.9	522.7	634.4		...	...	...	...
353	Steam and air conditioning supply		121.7	136.5	158.9	169.5		...	...	...	...
D	Electricity, gas, steam & air conditioning		5609.1	6014.0	6229.3	6297.0	a/	1400.6	1500.2	1481.9	1498.2
360	Water collection, treatment and supply		105.7	118.3	131.5	145.9		...	...	...	...
370	Sewerage		25.3	32.0	39.9	45.1		...	...	...	...
381	Waste collection		...	...	...	...		...	...	...	...
382	Waste treatment and disposal		...	...	...	...		...	...	...	...
383	Materials recovery		315.9	371.2	397.3	406.8		...	...	...	...
390	Remediation activities		...	...	...	...		...	...	...	...
E	Water supply;sewerage,waste management		446.9	521.5	568.6	597.7	a/	...	...	...	...

a/ D includes E.

China, Hong Kong SAR

ISIC Revision 4		Number of establishments					Number of employees					Wages and salaries paid to employees				
			(number)					(number)					(millions of Hong Kong Dollars)			
ISIC	Industry	Note	2012	2013	2014	2015	Note	2012	2013	2014	2015	Note	2012	2013	2014	2015
05	Mining of coal and lignite		...	...	...	...		...	...	...	...		...	...	...	...
06	Extraction of crude petroleum & natural gas		...	...	...	...		...	...	...	...		...	...	...	...
07	Mining of metal ores		...	...	...	...		...	...	...	...		...	...	...	...
08	Other mining and quarrying		2	2	2	2		100	100	100	100		...	...	...	...
09	Mining support service activities		...	...	...	...		...	...	...	...		...	...	...	...
B	Mining and quarrying		2	2	2	2		100	100	100	100		...	...	...	...
35	Electricity, gas, steam & air conditioning		12	8	8	8		7700	7700	7800	7900		3975	4235	4451	4642
D	Electricity, gas, steam & air conditioning		12	8	8	8		7700	7700	7800	7900		3975	4235	4451	4642
36	Water collection, treatment and supply		...	...	...	...		...	...	...	...		...	...	...	...
37	Sewerage		...	...	...	...		...	...	...	...		...	...	...	...
38	Waste collection,treatment,disposal activities		...	...	...	...		...	...	...	...		...	...	...	...
39	Remediation activities		...	...	...	...		...	...	...	...		...	...	...	...
E	Water supply;sewerage,waste management	a/	302	321	334	324	a/	3100	3200	3100	3300	a/	604	645	647	711

a/ E excludes 36.

China, Hong Kong SAR

ISIC Revision 4		Output at producers' prices					Value added at producers' prices				
		Note	(millions of Hong Kong Dollars)				Note	(millions of Hong Kong Dollars)			
ISIC	Industry		2012	2013	2014	2015		2012	2013	2014	2015
05	Mining of coal and lignite		...	...	...	...		...	...	...	...
06	Extraction of crude petroleum & natural gas		...	...	...	...		...	...	...	...
07	Mining of metal ores		...	...	...	...		...	...	...	...
08	Other mining and quarrying		...	...	...	...		...	...	...	...
09	Mining support service activities		...	...	...	...		...	...	...	...
B	Mining and quarrying		...	...	...	...		...	...	...	...
35	Electricity, gas, steam & air conditioning		54696	54196	56184	55099		33352	32425	34166	32910
D	Electricity, gas, steam & air conditioning		54696	54196	56184	55099		33352	32425	34166	32910
36	Water collection, treatment and supply		...	...	...	...		...	...	...	...
37	Sewerage		...	...	...	...		...	...	...	...
38	Waste collection,treatment,disposal activities		...	...	...	...		...	...	...	...
39	Remediation activities		...	...	...	...		...	...	...	...
E	Water supply;sewerage,waste management	a/	2889	3061	3227	3305	a/	1175	1238	1268	1281

a/ E excludes 36.

China, Macao SAR

ISIC Revision 3		Number of establishments					Number of employees					Wages and salaries paid to employees				
		Note	(number)				Note	(number)				Note	(thousands of Macao Patacas)			
ISIC	Industry		2012	2013	2014	2015		2012	2013	2014	2015		2012	2013	2014	2015
101	Mining and agglomeration of hard coal		...	...	...	...		...	...	...	...		...	...	...	...
102	Mining and agglomeration of lignite		...	...	...	...		...	...	...	...		...	...	...	...
103	Extraction and agglomeration of peat		...	...	...	...		...	...	...	...		...	...	...	...
111	Extraction of crude oil and natural gas		...	...	...	...		...	...	...	...		...	...	...	...
112	Oil, gas service activities		...	...	...	...		...	...	...	...		...	...	...	...
120	Mining of uranium and thorium ores		...	...	...	...		...	...	...	...		...	...	...	...
131	Iron ores		...	...	...	...		...	...	...	...		...	...	...	...
132	Non-ferrous metal ores,excl.uranium,thorium		...	...	...	...		...	...	...	...		...	...	...	...
141	Quarrying of stone, sand and clay		1	1	-	-		23	9	-	-		7105	3101	-	-
142	Mining and quarrying n.e.c.		...	...	...	...		...	...	...	...		...	...	...	...
C	Total mining and quarrying		1	1	-	-		23	9	-	-		7105	3101	-	-
401	Electricity production, collection, distribution		1	1	1	1		...	725	721	714		...	459867	570250	464705
402	Gas production and distribution		-	2	2	2		-	...	...	...		-	...	...	...
403	Steam and hot water supply		2	2	2	2		...	...	...	...		...	...	...	...
410	Water collection, purification, distribution		1	1	1	1		263	259	261	250		70110	88029	96172	100782
E	Total electricity, gas and water supply		4	6	6	6		1020	1097	1097	1080		373511	567486	689481	589628

China, Macao SAR

ISIC Revision 3		Output at producers' prices					Value added at producers' prices				
			(millions of Macao Patacas)					(millions of Macao Patacas)			
ISIC	Industry	Note	2012	2013	2014	2015	Note	2012	2013	2014	2015
101	Mining and agglomeration of hard coal		...	...	...	...		...	...	...	...
102	Mining and agglomeration of lignite		...	...	...	...		...	...	...	...
103	Extraction and agglomeration of peat		...	...	...	...		...	...	...	...
111	Extraction of crude oil and natural gas		...	...	...	...		...	...	...	...
112	Oil, gas service activities		...	...	...	...		...	...	...	...
120	Mining of uranium and thorium ores		...	...	...	...		...	...	...	...
131	Iron ores		...	...	...	...		...	...	...	...
132	Non-ferrous metal ores,excl.uranium,thorium		...	...	...	...		...	...	...	...
141	Quarrying of stone, sand and clay		35.0	4.3	-	-		17.2	3.0	-	-
142	Mining and quarrying n.e.c.		...	...	...	...		...	...	...	...
C	Total mining and quarrying		35.0	4.3	-	-		17.2	3.0	-	-
401	Electricity production, collection, distribution		...	2178.0	2447.5	2788.0		...	1737.9	2002.1	2105.6
402	Gas production and distribution		-	...	...	...		-	...	...	...
403	Steam and hot water supply		...	...	...	...		...	...	...	...
410	Water collection, purification, distribution		444.8	466.6	480.1	554.2		199.2	222.8	233.6	265.2
E	Total electricity, gas and water supply		2693.3	2663.1	3105.2	3364.9		1820.3	1962.7	2240.1	2373.1

China, Taiwan Province

ISIC Revision 4		Establishments					Number of employees					Wages and salaries paid to employees				
			(number)					(number)					(millions of Taiwan Dollars)			
ISIC	Industry	Note	2012	2013	2014	2015	Note	2012	2013	2014	2015	Note	2012	2013	2014	2015
051	Mining of hard coal		...	...	...	...		...	...	...	...		...	...	...	...
052	Mining of lignite		...	...	...	...		...	...	...	...		...	...	...	...
061	Extraction of crude petroleum		...	...	...	...		1301a/	...	1413a/	...		1486.7a/	...	1564.8a/	...
062	Extraction of natural gas		...	...	...	...		...a/	...	...a/	...		...a/	...	...a/	...
071	Mining of iron ores		...	...	...	...		...	...	...	...		...	...	...	...
072	Mining of non-ferrous metal ores		...	...	...	...		...	...	...	...		...	...	...	...
081	Quarrying of stone, sand and clay		...	...	...	...		2688	...	2495	...		1056.5	...	995.0	...
089	Mining and quarrying n.e.c.		...	...	...	...		...	...	...	...		...	...	...	...
091	Support activities for petroleum		...	...	...	...		...	...	...	...		...	...	...	...
099	Support activities for other mining, quarrying		...	...	...	...		...	...	...	...		...	...	...	...
B	Mining and quarrying		...	...	...	...		4156	3872	3908	...		2630.8	5145.0	5313.0	...
351	Electric power generation, transmission		...	...	...	...		23881	...	24037	...		28077.5	37588.0	37475.0	...
352	Manufacture of gas		...	...	...	...		5519	...	5074	...		4541.3	4903.0	5109.0	...
353	Steam and air conditioning supply		...	...	...	...		...	...	...	...		...	...	...	...
D	Electricity, gas, steam & air conditioning		...	...	...	...		29000	29000	...	...		...	42491.0	42584.0	...
360	Water collection, treatment and supply		...	...	...	...		6223	...	6189	...		6190.4	10698.0	10740.0	...
370	Sewerage		...	...	...	...		1720	...	1923	...		690.3	15889.0b/	16552.0b/	...
381	Waste collection		...	...	...	...		7323	...	8333	...		2575.2	...b/	...b/	...
382	Waste treatment and disposal		...	...	...	...		5141	...	5722	...		2244.8	...b/	...b/	...
383	Materials recovery		...	...	...	...		...	...	...	...		...	...b/	...b/	...
390	Remediation activities		...	...	...	...		5705	...	5662	...		2096.2	...b/	...b/	...
E	Water supply;sewerage,waste management		...	...	...	...		82000	84000	82000	...		...	26587.0	27292.0	...

a/ 061 includes 062.
b/ 370 includes 381, 382, 383 and 390.

China, Taiwan Province

ISIC Revision 4		Output (valuation not defined)					Value added (valuation not defined)				
			(millions of Taiwan Dollars)					(millions of Taiwan Dollars)			
ISIC	Industry	Note	2012	2013	2014	2015	Note	2012	2013	2014	2015
051	Mining of hard coal		...	...	...	...		...	...	...	...
052	Mining of lignite		...	...	...	...		...	...	...	...
061	Extraction of crude petroleum		...	...	...	...		...	...	...	...
062	Extraction of natural gas		...	...	...	...		...	...	...	...
071	Mining of iron ores		...	...	...	...		...	...	...	...
072	Mining of non-ferrous metal ores		...	...	...	...		...	...	...	...
081	Quarrying of stone, sand and clay		...	...	...	...		...	...	...	...
089	Mining and quarrying n.e.c.		...	...	...	...		...	...	...	...
091	Support activities for petroleum		...	...	...	...		...	...	...	...
099	Support activities for other mining, quarrying		...	...	...	...		...	...	...	...
B	Mining and quarrying		28510	28541	30701	...		15500	15471	16766	16597
351	Electric power generation, transmission		590591	625205	686600	...		123251	174178	208454	...
352	Manufacture of gas		105474	124112	133898	...		11644	28114	33209	...
353	Steam and air conditioning supply		...	...	...	...		...	...	...	...
D	Electricity, gas, steam & air conditioning		696065	749317	820498	...		134895	202292	242194	307752
360	Water collection, treatment and supply		42628	43922	44936	...		22489	22659	22096	...
370	Sewerage		168826a/	169932a/	170359a/	...		78448a/	77825a/	78000a/	...
381	Waste collection		...a/	...a/	...a/	...		...a/	...a/	...a/	...
382	Waste treatment and disposal		...a/	...a/	...a/	...		...a/	...a/	...a/	...
383	Materials recovery		...a/	...a/	...a/	...		...a/	...a/	...a/	...
390	Remediation activities		...a/	...a/	...a/	...		...a/	...a/	...a/	...
E	Water supply;sewerage,waste management		211454	213854	215295	...		100937	100484	101907	102261

a/ 370 includes 381, 382, 383 and 390.

Croatia

ISIC Revision 4		Number of enterprises					Number of employees					Wages and salaries paid to employees				
			(number)					(number)					(thousands of Croatian Kunas)			
ISIC	Industry	Note	2012	2013	2014	2015	Note	2012	2013	2014	2015	Note	2012	2013	2014	2015
051	Mining of hard coal		-	-	-	-		-	-	-	-		-	-	-	-
052	Mining of lignite		-	-	-	-		-	-	-	-		-	-	-	-
061	Extraction of crude petroleum		-	-	-	-		-	-	-	-		-	-	-	-
062	Extraction of natural gas		4	4	4	4		...	...	...	7852		...	...	...	...
071	Mining of iron ores		-	-	-	-		-	-	-	-		-	-	-	-
072	Mining of non-ferrous metal ores		-	-	-	-		-	-	-	-		-	-	-	-
081	Quarrying of stone, sand and clay		243	224	212	205		2156	2128	2016	1877		161867	167095	155465	149471
089	Mining and quarrying n.e.c.		19	18	18	16		147	149	...	149		12628	12150	...	...
091	Support activities for petroleum		8	6	4	4		...	...	2556	2272		...	...	459439	396166a/
099	Support activities for other mining, quarrying		7	5	7	4		...	...	...	31		...	...	...	...a/
B	Mining and quarrying		281	257	245	233		14132	13685	13181	12181		1992268	1904183	1849491	1787136
351	Electric power generation, transmission		307	432	477	490		13275	12202	11999	12149		1690929	1651400	1723280	1629974
352	Manufacture of gas		57	58	64	64		1782	1787	1744	1754		208711	204385	203221	208369
353	Steam and air conditioning supply		24	23	19	19		1176	1142	728	718		118380	114478	82184	97189
D	Electricity, gas, steam & air conditioning		388	513	560	573		16233	15131	14471	14621		2018020	1970262	2008686	1935532
360	Water collection, treatment and supply		125	119	146	151		9268	9457	8463	8530		829065	750783	822739	822175
370	Sewerage		59	59	66	69		338	541	543	557		30008	51224	57259	54872
381	Waste collection		269	286	320	314		7802	8728	9307	9592		639698	783910	793088	798912
382	Waste treatment and disposal		58	80	100	115		449	892	992	1138		28307	89684	67277	82711
383	Materials recovery		149	137	120	110		1870	1854	1784	1745		175493	165430	153839	143965
390	Remediation activities		84	85	93	94		1376	1483	1518	1561		160357	163301	172615	177835
E	Water supply;sewerage,waste management		744	766	845	853		21103	22955	22607	23123		1862928	2004331	2066816	2080470

a/ 091 includes 099.

Croatia

ISIC Revision 4		Output (valuation not defined)					Value added at factor values				
		Note	(millions of Croatian Kunas)				Note	(millions of Croatian Kunas)			
ISIC	Industry		2012	2013	2014	2015		2012	2013	2014	2015
051	Mining of hard coal		-	-	-	-		-	-	-	-
052	Mining of lignite		-	-	-	-		-	-	-	-
061	Extraction of crude petroleum		-	-	-	-		-	-	-	-
062	Extraction of natural gas		...	...	...	...		...	...	...	...
071	Mining of iron ores		-	-	-	-		-	-	-	-
072	Mining of non-ferrous metal ores		-	-	-	-		-	-	-	-
081	Quarrying of stone, sand and clay		1031.8	1100.9	1222.4	1047.0		401.9	355.3	543.1	426.2
089	Mining and quarrying n.e.c.		95.8	114.4	...	...		30.7	38.5	...	...
091	Support activities for petroleum		...	...	2074.5	1964.9a/		...	...	906.8	636.1a/
099	Support activities for other mining, quarrying		...	...	...	...a/		...	...	...	...a/
B	Mining and quarrying		29735.9	28247.6	25633.9	19453.9		9994.2	9273.6	8761.1	6242.8
351	Electric power generation, transmission		26893.4	27429.6	26041.7	24455.6		5569.4	7473.7	8311.8	7614.9
352	Manufacture of gas		3964.0	4983.9	6567.2	7747.0		644.8	900.9	779.7	1114.0
353	Steam and air conditioning supply		967.3	1132.6	935.8	946.9		-189.0	132.8	144.2	244.1
D	Electricity, gas, steam & air conditioning		31824.7	33546.2	33544.6	33149.5		6025.1	8507.5	9235.7	8972.9
360	Water collection, treatment and supply		2676.4	2410.0	2971.9	2971.2		1744.4	1565.7	2108.9	2113.9
370	Sewerage		158.1	216.9	254.3	265.3		94.1	124.4	165.3	174.7
381	Waste collection		1775.9	2100.0	2184.3	2230.5		1082.6	1397.6	1290.7	1297.1
382	Waste treatment and disposal		129.8	312.2	240.3	304.3		67.9	165.4	121.0	171.2
383	Materials recovery		1680.3	1566.0	1329.0	1083.8		443.4	386.8	346.9	363.6
390	Remediation activities		450.0	408.2	512.5	541.6		239.0	232.2	280.3	295.8
E	Water supply;sewerage,waste management		6870.6	7013.3	7492.3	7396.7		3671.5	3872.0	4313.1	4416.3

a/ 091 includes 099.

Curaçao

ISIC Revision 4		Number of enterprises					Employees					Wages and salaries				
			(number)					(number)					(millions of Netherlands Antillean Guilders)			
ISIC	Industry	Note	2012	2013	2014	2015	Note	2012	2013	2014	2015	Note	2012	2013	2014	2015
051	Mining of hard coal		...	...	...	...		...	...	...	...		...	...	...	...
052	Mining of lignite		...	...	...	...		...	...	...	...		...	...	...	...
061	Extraction of crude petroleum		...	...	...	...		...	...	...	...		...	...	...	...
062	Extraction of natural gas		...	...	...	...		...	...	...	...		...	...	...	...
071	Mining of iron ores		...	...	...	...		...	...	...	...		...	...	...	...
072	Mining of non-ferrous metal ores		...	...	...	...		...	...	...	...		...	...	...	...
081	Quarrying of stone, sand and clay		...	...	1	2		...	...	...	...		...	...	...	...
089	Mining and quarrying n.e.c.		...	...	...	...		...	...	...	...		...	...	...	...
091	Support activities for petroleum		...	...	1	...		...	...	...	...		...	...	...	...
099	Support activities for other mining, quarrying		...	...	...	...		...	...	...	...		...	...	...	...
B	Mining and quarrying		...	...	2a/	2a/		...	...	...	...		...	...	...	...
351	Electric power generation, transmission		...	...	2	1		...	...	...	...		...	...	...	...
352	Manufacture of gas		...	...	1	1		...	...	...	...		...	...	...	...
353	Steam and air conditioning supply		...	...	...	...		...	...	...	...		...	...	...	...
D	Electricity, gas, steam & air conditioning		...	...	3a/	2a/		...	...	...	...		...	...	...	...
360	Water collection, treatment and supply		...	...	3	1		...	...	...	...		...	...	...	...
370	Sewerage		...	...	6	5		...	...	...	...		...	...	...	...
381	Waste collection		...	...	1	2		...	...	...	...		...	...	...	...
382	Waste treatment and disposal		...	...	1	1		...	...	...	...		...	...	...	...
383	Materials recovery		...	...	15	7		...	...	...	...		...	...	...	...
390	Remediation activities		...	...	...	...		...	...	...	...		...	...	...	...
E	Water supply;sewerage,waste management		...	...	26a/	16a/		...	...	...	...		...	...	...	...

a/ Sum of available data.

Cyprus

ISIC Revision 4		Number of enterprises					Number of employees					Wages and salaries paid to employees				
			(number)					(number)					(thousands of Euros)			
ISIC	Industry	Note	2012	2013	2014	2015	Note	2012	2013	2014	2015	Note	2012	2013	2014	2015
051	Mining of hard coal		-	-	-	-		-	-	-	-		-	-	-	-
052	Mining of lignite		-	-	-	-		-	-	-	-		-	-	-	-
061	Extraction of crude petroleum		-	-	-	-		-	-	-	-		-	-	-	-
062	Extraction of natural gas		-	-	-	-		-	-	-	-		-	-	-	-
071	Mining of iron ores		-	-	-	-		-	-	-	-		-	-	-	-
072	Mining of non-ferrous metal ores		-	-	-	-		-	-	-	-		-	-	-	-
081	Quarrying of stone, sand and clay		...	...	...	...		...	...	...	...		...	...	...	...
089	Mining and quarrying n.e.c.		...	...	...	...		...	...	...	...		...	...	...	...
091	Support activities for petroleum		...	...	...	...		...	...	...	...		...	...	...	...
099	Support activities for other mining, quarrying		-	-	-	-		-	-	-	-		-	-	-	-
B	Mining and quarrying		92a/	108a/	117a/	58b/		2882a/	2744a/	2621a/	420b/		97152a/	91804a/	...	12072b/
351	Electric power generation, transmission		...	...	...	58		...	...	...	2115		...	...	...	70970
352	Manufacture of gas		-	-	-	-		-	-	-	-		-	-	-	-
353	Steam and air conditioning supply		-	-	-	-		-	-	-	-		-	-	-	-
D	Electricity, gas, steam & air conditioning		...a/	...a/	...a/	58		...a/	...a/	...a/	2115		...a/	...a/	...	70970
360	Water collection, treatment and supply		9	7	5	6		392	361	355	376		11514	10008	10003	9830
370	Sewerage		66	66	61	60		409	405	377	383		12202	10917	10501	9869
381	Waste collection		74	73	72c/	68		370	377	360c/	380		6782	6179	6188c/	5661
382	Waste treatment and disposal		11	12	11	10d/		71	101	95	99d/		1323	1503	1526	1552d/
383	Materials recovery		38	44e/	45	40		379	391e/	405	395		6076	6159e/	6356	5822
390	Remediation activities		-	...e/	...c/	...d/		-	...e/	...c/	...d/		-	...e/	...c/	...d/
E	Water supply;sewerage,waste management		198	202	194	184		1621	1635	1592	1628		37897	34766	34574	32883

a/ B includes D.
b/ B includes 08 and 09 only.
c/ 381 includes 390.
d/ 382 includes 390.
e/ 383 includes 390.

Cyprus

ISIC Revision 4		Output at producers' prices					Value added at producers' prices				
			(thousands of Euros)					(thousands of Euros)			
ISIC	Industry	Note	2012	2013	2014	2015	Note	2012	2013	2014	2015
051	Mining of hard coal		-	-	-	-		-	-	-	-
052	Mining of lignite		-	-	-	-		-	-	-	-
061	Extraction of crude petroleum		-	-	-	-		-	-	-	-
062	Extraction of natural gas		-	-	-	-		-	-	-	-
071	Mining of iron ores		-	-	-	-		-	-	-	-
072	Mining of non-ferrous metal ores		-	-	-	-		-	-	-	-
081	Quarrying of stone, sand and clay		...	...	...	...		...	...	...	...
089	Mining and quarrying n.e.c.		...	...	...	...		...	...	...	...
091	Support activities for petroleum		...	...	...	...		...	...	...	...
099	Support activities for other mining, quarrying		-	-	-	-		-	-	-	-
B	Mining and quarrying		1197075a/	990095a/	847220a/	98371b/		398283a/	358036a/	284909a/	28945b/
351	Electric power generation, transmission		...	...	...	659907		...	...	...	...
352	Manufacture of gas		-	-	-	-		-	-	-	-
353	Steam and air conditioning supply		-	-	-	-		-	-	-	-
D	Electricity, gas, steam & air conditioning		...a/	...a/	...a/	659907		...a/	...a/	...a/	...
360	Water collection, treatment and supply		85283	80519	79897	84459		37385	38717	34039	...
370	Sewerage		103002	104184	102297	100196		66314	74319	74069	...
381	Waste collection		24336	19882	19750c/	17590		13980	11517	10941c/	...
382	Waste treatment and disposal		9530	9401	10607	14080d/		4966	4731	4808	...
383	Materials recovery		102268	76967e/	68558	55810		19673	16541e/	16134	...
390	Remediation activities		-	...e/	...c/	...d/		-	...e/	...c/	...
E	Water supply;sewerage,waste management		324419	290953	281109	272135		142318	145825	139991	...

a/ B includes D.
b/ B includes 08 and 09 only.
c/ 381 includes 390.
d/ 382 includes 390.
e/ 383 includes 390.

Czechia

ISIC Revision 4		Number of enterprises					Number of employees					Wages and salaries paid to employees				
			(number)					(thousands)					(millions of Czech Koruny)			
ISIC	Industry	Note	2012	2013	2014	2015	Note	2012	2013	2014	2015	Note	2012	2013	2014	2015
051	Mining of hard coal		5	6	5	6	a/	22.4	21.5	19.9	18.6	a/	10110	9140	8330	7769
052	Mining of lignite		6	6	6	6	a/	...	...	...	...	a/	...	...	...	...
061	Extraction of crude petroleum		3	3	3	3		...	...	...	...		...	...	...	...
062	Extraction of natural gas		2	2	2	2		...	...	...	...		...	...	...	...
071	Mining of iron ores		-	-	-	-		-	-	-	-		-	-	-	-
072	Mining of non-ferrous metal ores		1	1	1	1		...	...	...	...		...	...	...	...
081	Quarrying of stone, sand and clay		270	271	290	280		5.4	5.2	4.9	5.0		1781	1665	1649	1721
089	Mining and quarrying n.e.c.		37	41	43	38		0.5	0.5	0.2	0.2		166	153	80	76
091	Support activities for petroleum		5	5	5	6		0.5	2.3b/	0.6	2.2b/		206	746b/	234	769b/
099	Support activities for other mining, quarrying		30	28	25	34		1.7	...b/	1.6	...b/		546	...b/	452	...b/
B	Mining and quarrying		359	363	380	376		33.6	32.6	30.4	29.2		13993	12904	12008	11637
351	Electric power generation, transmission		5546	8000	9969	10544		15.3	15.8	16.3	16.5		9577	9584	9836	9640
352	Manufacture of gas		41	46	47	48		2.6	2.1	1.5	1.5		1335	948	859	807
353	Steam and air conditioning supply		404	400	398	404		12.4	11.7	11.7	11.7		4854	4484	4409	4588
D	Electricity, gas, steam & air conditioning		5991	8446	10414	10996		30.3	29.6	29.4	29.7		15766	15016	15103	15036
360	Water collection, treatment and supply		446	472	530	559		17.7	17.6	17.7	17.6		5850	5996	6185	6239
370	Sewerage		349	312	301	282		1.2	1.1	1.0	1.1		360	348	330	338
381	Waste collection		4120	4049	4121	4110		20.4	20.7	20.5	20.8		5382	5341	5411	5685
382	Waste treatment and disposal		406	386	365	359		2.9	2.8	2.8	2.5		890	873	856	769
383	Materials recovery		982	900	843	789		5.1	4.9	4.6	4.5		1268	1200	1190	1200
390	Remediation activities		80	77	71	65		0.6	0.6	0.6	0.7		231	226	231	248
E	Water supply;sewerage,waste management		6383	6196	6231	6164		47.8	47.7	47.2	47.1		13983	13985	14200	14480

a/ 051 includes 052.
b/ 091 includes 099.

Czechia

ISIC Revision 4		Output (valuation not defined)					Value added at factor values				
			(millions of Czech Koruny)					(millions of Czech Koruny)			
ISIC	Industry	Note	2012	2013	2014	2015	Note	2012	2013	2014	2015
051	Mining of hard coal	a/	61082	50578	46464	42924	a/	29638	22649	20627	19859
052	Mining of lignite	a/	...	...	...	...	a/	...	...	...	...
061	Extraction of crude petroleum		...	...	...	...		...	...	...	...
062	Extraction of natural gas		...	...	...	...		...	...	...	...
071	Mining of iron ores		-	-	-	-		-	-	-	-
072	Mining of non-ferrous metal ores		...	...	...	...		...	...	...	...
081	Quarrying of stone, sand and clay		13970	13736	14500	15917		4424	4380	4923	5546
089	Mining and quarrying n.e.c.		1728	2006	856	927		438	473	207	243
091	Support activities for petroleum	b/	4487	4458	4026	3792	b/	1788	2065	1652	1631
099	Support activities for other mining, quarrying	b/	...	...	...	...	b/	...	...	...	...
B	Mining and quarrying		86741	76496	72004	68552		40299	33631	31628	30757
351	Electric power generation, transmission		682300	666587	611230	632082		122194	128775	111199	114926
352	Manufacture of gas		271838	270241	218561	215149		-4102	7506	10742	11924
353	Steam and air conditioning supply		60742	58575	52316	53115		17934	17692	16188	16809
D	Electricity, gas, steam & air conditioning		1014880	995403	882107	900346		136026	153973	138129	143659
360	Water collection, treatment and supply		36464	37855	39539	40954		15034	15357	15762	16395
370	Sewerage		2447	2359	2175	2087		956	977	911	996
381	Waste collection		40382	38762	38663	39249		12753	12616	13198	13847
382	Waste treatment and disposal		5347	5534	5689	5439		2133	2091	2123	1899
383	Materials recovery		24940	22190	22990	20582		3259	2642	2924	2698
390	Remediation activities		1262	1289	1779	1222		103	99	936	562
E	Water supply;sewerage,waste management		110842	107988	110835	109533		34240	33782	35857	36396

a/ 051 includes 052.
b/ 091 includes 099.

Denmark

ISIC Revision 4		Number of enterprises					Number of employees					Wages and salaries paid to employees				
			(number)					(number)					(millions of Danish Kroner)			
ISIC	Industry	Note	2012	2013	2014	2015	Note	2012	2013	2014	2015	Note	2012	2013	2014	2015
051	Mining of hard coal		-	-	-	-		-	-	-	-		-	-	-	-
052	Mining of lignite		-	-	-	-		-	-	-	-		-	-	-	-
061	Extraction of crude petroleum		9	10	11	12		952	966	1024	1046		552	569	724	867
062	Extraction of natural gas		-	-	-	-		-	-	-	-		-	-	-	-
071	Mining of iron ores		-	-	-	-		-	-	-	-		-	-	-	-
072	Mining of non-ferrous metal ores		-	-	-	-		-	-	-	-		-	-	-	-
081	Quarrying of stone, sand and clay		136	129	120	127		825	772	779	805		269	262	268	293
089	Mining and quarrying n.e.c.		13	10	12	10		470	459	454	451		162	160	158	163
091	Support activities for petroleum		49	52	55	55		3434	1966	1932	1535		1583	873	888	671
099	Support activities for other mining, quarrying		11	9	9	9		77	133	123	77		33	46	47	28
B	Mining and quarrying		218	210	207	213		5758	4296	4312	3914		2599	1909	2084	2022
351	Electric power generation, transmission		1447	1387	1331	1285		7453	8782	9903	9448		2471	4226	4152	4062
352	Manufacture of gas		54	61	57	62		1851	569	...	887		1869	509	...	356
353	Steam and air conditioning supply		401	397	396	398		1453	1499	...	1896		616	676	...	713
D	Electricity, gas, steam & air conditioning		1902	1845	1784	1745		10757	10850	12640	12231		4955	5410	5232	5131
360	Water collection, treatment and supply		2012	2007	2007	1975		2546	2481	...	...		345	212	255	263
370	Sewerage		241	254	252	263		2101	1885	2177	2293		851	784	751	790
381	Waste collection		113	104	102	97		2073	2198	4177	4319		1246	923	1328	1355
382	Waste treatment and disposal		69	69	65	65		1959	2034	2575	2697		809	841	839	870
383	Materials recovery		147	127	137	132		1698	1551	1689	1788		529	494	563	583
390	Remediation activities		18	18	20	18		53	80	78	86		36	7	22	24
E	Water supply;sewerage,waste management		2600	2579	2583	2550		10430	10229	11717	12195		3816	3261	3760	3884

Denmark

ISIC Revision 4		Output (valuation not defined)					Value added at factor values				
			(millions of Danish Kroner)					(millions of Danish Kroner)			
ISIC	Industry	Note	2012	2013	2014	2015	Note	2012	2013	2014	2015
051	Mining of hard coal		-	-	-	-		-	-	-	-
052	Mining of lignite		-	-	-	-		-	-	-	-
061	Extraction of crude petroleum		54956	51202	43461	44182		43398	38801	28920	29061
062	Extraction of natural gas		-	-	-	-		-	-	-	-
071	Mining of iron ores		-	-	-	-		-	-	-	-
072	Mining of non-ferrous metal ores		-	-	-	-		-	-	-	-
081	Quarrying of stone, sand and clay		1976	1768	1658	2129		630	548	587	767
089	Mining and quarrying n.e.c.		799	775	536	907		334	315	306	351
091	Support activities for petroleum		4697	2932	3004	4512		2650	1739	1984	1648
099	Support activities for other mining, quarrying		93	94	82	68		28	44	42	31
B	Mining and quarrying		62521	56771	48739	51798		47040	41447	31839	31858
351	Electric power generation, transmission		52346	104695	65719	73558		14788	20080	17575	18725
352	Manufacture of gas		63624	46903	...	4026		-2696	69	...	2000
353	Steam and air conditioning supply		16220	17816	...	20134		3932	3531	...	4133
D	Electricity, gas, steam & air conditioning		132190	169415	137411	97719		16023	23679	21866	24858
360	Water collection, treatment and supply		6261	6196	6738	7152		2791	2721	2816	2997
370	Sewerage		9877	11159	10168	10297		5488	6567	5879	5766
381	Waste collection		5476	4911	5695	5634		1998	1407	2055	2128
382	Waste treatment and disposal		5572	6392	5937	5765		2022	1909	2343	2351
383	Materials recovery		5825	5480	6430	5559		986	842	972	1027
390	Remediation activities		51	134	178	173		-7	35	37	43
E	Water supply;sewerage,waste management		33060	34273	35146	34581		13278	13480	14104	14313

Ecuador

ISIC Revision 4		Number of establishments					Number of employees					Wages and salaries paid to employees				
			(number)					(number)					(thousands of US Dollars)			
ISIC	Industry	Note	2012	2013	2014	2015	Note	2012	2013	2014	2015	Note	2012	2013	2014	2015
051	Mining of hard coal		...	...	...	...		...	...	...	...		...	...	...	...
052	Mining of lignite		...	...	...	...		...	...	...	...		...	...	...	...
061	Extraction of crude petroleum		...	...	...	...		3796	5844	6556	6734		59083	121438	141817	126411
062	Extraction of natural gas		...	...	...	...		...	...	...	...		...	...	...	...
071	Mining of iron ores		...	...	...	...		...	...	...	...		...	...	...	...
072	Mining of non-ferrous metal ores		41	25	25	23		3892	4697	4873	4978		23837	32243	38289	40950
081	Quarrying of stone, sand and clay		11	8	9	8		368	227	309	270		2843	1801	3002	2645
089	Mining and quarrying n.e.c.		...	...	...	...		141	166	...	78		1086	1267	...	342
091	Support activities for petroleum		...	...	...	6		...	...	...	755		...	...	...	19339
099	Support activities for other mining, quarrying		...	...	...	...		...	...	...	31		...	...	...	1678
B	Mining and quarrying		55	35	35	40		8197	10934	11738	12846		86848	156750	183108	191365
351	Electric power generation, transmission		...	...	...	...		...	...	...	...		...	...	...	...
352	Manufacture of gas		...	...	...	...		...	...	...	...		...	...	...	...
353	Steam and air conditioning supply		...	...	...	...		...	...	...	...		...	...	...	...
D	Electricity, gas, steam & air conditioning		...	...	...	...		...	...	...	...		...	...	...	...
360	Water collection, treatment and supply		...	...	...	...		...	...	...	...		...	...	...	...
370	Sewerage		...	...	...	...		...	...	...	...		...	...	...	...
381	Waste collection		...	...	...	...		...	...	...	...		...	...	...	...
382	Waste treatment and disposal		...	...	...	...		...	...	...	...		...	...	...	...
383	Materials recovery		...	...	...	...		...	...	...	...		...	...	...	...
390	Remediation activities		...	...	...	...		...	...	...	...		...	...	...	...
E	Water supply;sewerage,waste management		...	...	...	...		...	...	...	...		...	...	...	...

Ecuador

ISIC Revision 4		Output at basic prices					Value added at basic prices				
			(thousands of US Dollars)					(thousands of US Dollars)			
ISIC	Industry	Note	2012	2013	2014	2015	Note	2012	2013	2014	2015
051	Mining of hard coal		...	...	...	...		...	...	...	...
052	Mining of lignite		...	...	...	...		...	...	...	...
061	Extraction of crude petroleum		5758974	6716904	6932299	5606433		4640207	4662081	4875895	-21572
062	Extraction of natural gas		...	...	...	...		...	...	...	...
071	Mining of iron ores		...	...	...	...		...	...	...	...
072	Mining of non-ferrous metal ores		118860	152937	160837	182431		70813	80863	87838	95808
081	Quarrying of stone, sand and clay		71152	15936	33613	26085		50715	5854	10660	7255
089	Mining and quarrying n.e.c.		5670	5276	...	870		2291	3684	...	568
091	Support activities for petroleum		...	...	...	368575		...	...	...	14279
099	Support activities for other mining, quarrying		...	...	...	4611		...	...	...	1822
B	Mining and quarrying		5954655	6891053	7126748	6189005		4764026	4752481	4974392	98159
351	Electric power generation, transmission		...	...	...	...		...	...	...	...
352	Manufacture of gas		...	...	...	...		...	...	...	...
353	Steam and air conditioning supply		...	...	...	...		...	...	...	...
D	Electricity, gas, steam & air conditioning		...	...	...	...		...	...	...	...
360	Water collection, treatment and supply		...	...	...	...		...	...	...	...
370	Sewerage		...	...	...	...		...	...	...	...
381	Waste collection		...	...	...	...		...	...	...	...
382	Waste treatment and disposal		...	...	...	...		...	...	...	...
383	Materials recovery		...	...	...	...		...	...	...	...
390	Remediation activities		...	...	...	...		...	...	...	...
E	Water supply;sewerage,waste management		...	...	...	...		...	...	...	...

Estonia

ISIC Revision 4		Number of enterprises					Number of employees					Wages and salaries paid to employees				
			(number)					(number)					(millions of Euros)			
ISIC	Industry	Note	2012	2013	2014	2015	Note	2012	2013	2014	2015	Note	2012	2013	2014	2015
051	Mining of hard coal		-	-	-	-		-	-	-	-		-	-	-	-
052	Mining of lignite		-	-	-	-		-	-	-	-		-	-	-	-
061	Extraction of crude petroleum		2	2	2	2		...	...	...	...		...	...	...	...
062	Extraction of natural gas		-	-	-	-		-	-	-	-		-	-	-	-
071	Mining of iron ores		-	-	-	-		-	-	-	-		-	-	-	-
072	Mining of non-ferrous metal ores		-	-	-	-		-	-	-	-		-	-	-	-
081	Quarrying of stone, sand and clay		69	83	88	86		...	...	...	695		...	...	...	10.5
089	Mining and quarrying n.e.c.		44	42	41	45		1058	1072	1036	966		11.6	13.1	13.0	12.6
091	Support activities for petroleum		-	-	-	-		-	-	-	-		-	-	-	-
099	Support activities for other mining, quarrying		13	11	11	10		31	19	35	...		0.3	0.2	0.5	...
B	Mining and quarrying		128	138	142	143		5267	5073	4832	4742		70.9	74.1	75.6	76.4
351	Electric power generation, transmission		101	99	97	101		3363	3244	3142	3038		60.9	61.5	60.4	59.2
352	Manufacture of gas		17	18	25	23		303	315	314	331		5.4	6.4	6.8	6.3
353	Steam and air conditioning supply		113	109	104	106		1725	1616	1588	1521		19.4	19.6	20.2	20.9
D	Electricity, gas, steam & air conditioning		231	226	226	230		5391	5175	5044	4890		85.7	87.4	87.4	86.4
360	Water collection, treatment and supply		74	77	81	77		1252	1196	1173	1133		14.4	14.6	15.2	16.0
370	Sewerage		36	45	46	47		176	...	180	186		1.8	...	2.2	2.1
381	Waste collection		54	51	48	45		1139	1115	1108	1172		12.6	13.6	14.1	14.8
382	Waste treatment and disposal		28	29	27	28		300	287	346	397		3.3	3.5	4.1	4.4
383	Materials recovery		79	82	70	76		428	421	299	333		3.5	3.3	2.4	3.0
390	Remediation activities		11	8	8	9		30	...	17	22		0.2	...	0.2	0.2
E	Water supply;sewerage,waste management		282	292	280	282		3325	3243	3123	3243		35.8	37.1	38.2	40.3

Estonia

ISIC Revision 4		Output at producers' prices					Value added at factor values				
			(millions of Euros)					(millions of Euros)			
ISIC	Industry	Note	2012	2013	2014	2015	Note	2012	2013	2014	2015
051	Mining of hard coal		-	-	-	-		-	-	-	-
052	Mining of lignite		-	-	-	-		-	-	-	-
061	Extraction of crude petroleum		...	...	...	...		...	...	...	...
062	Extraction of natural gas		-	-	-	-		-	-	-	-
071	Mining of iron ores		-	-	-	-		-	-	-	-
072	Mining of non-ferrous metal ores		-	-	-	-		-	-	-	-
081	Quarrying of stone, sand and clay		...	...	...	59.7		...	...	...	26.7
089	Mining and quarrying n.e.c.		74.6	79.6	82.9	75.2		30.1	34.3	34.1	30.8
091	Support activities for petroleum		-	-	-	-		-	-	-	-
099	Support activities for other mining, quarrying		1.4	1.2	2.1	...		0.7	0.6	1.1	...
B	Mining and quarrying		359.4	410.4	415.7	397.0		154.0	211.0	213.4	187.8
351	Electric power generation, transmission		677.5	606.4	521.4	422.7		383.2	435.8	480.7	393.8
352	Manufacture of gas		...	...	...	...		30.9	33.8	28.3	21.9
353	Steam and air conditioning supply		258.7	249.5	243.5	218.0		109.5	127.0	117.2	111.3
D	Electricity, gas, steam & air conditioning		936.3	856.5	765.0	640.7		523.7	596.6	626.2	526.9
360	Water collection, treatment and supply		...	...	...	...		70.4	68.2	67.8	71.1
370	Sewerage		...	...	...	...		4.5	...	4.2	3.8
381	Waste collection		...	...	...	...		28.1	31.0	33.3	31.8
382	Waste treatment and disposal		...	...	...	...		8.8	9.3	12.7	12.0
383	Materials recovery		...	...	...	...		10.4	1.7	7.0	11.2
390	Remediation activities		...	...	...	...		0.6	...	0.3	0.5
E	Water supply;sewerage,waste management		...	...	...	...		122.7	114.5	125.4	130.5

Finland

ISIC Revision 4		Number of enterprises					Number of employees					Wages and salaries paid to employees			
		Note	(number)				Note	(number)				Note	(millions of Euros)		
ISIC	Industry		2012	2013	2014	2015		2012	2013	2014	2015		2012	2013	2014	2015
05	Mining of coal and lignite		-	-	-	-		-	-	-	-		-	-	-	-
051	Mining of hard coal		-	-	-	-		-	-	-	-		-	-	-	-
052	Mining of lignite		-	-	-	-		-	-	-	-		-	-	-	-
06	Extraction of crude petroleum & natural gas		-	-	-	-		-	-	-	-		-	-	-	-
061	Extraction of crude petroleum		-	-	-	-		-	-	-	-		-	-	-	-
062	Extraction of natural gas		-	-	-	-		-	-	-	-		-	-	-	-
07	Mining of metal ores		24	27	25	26		1911	2145	2064	1716		88	104	91	90
071	Mining of iron ores		-	1	1	1		-	2145a/	2064a/	1716a/		-	104a/	91a/	90a/
072	Mining of non-ferrous metal ores		24	26	24	25		1911	...a/	...a/	...a/		88	...a/	...a/	...a/
08	Other mining and quarrying		828	823	824	797		3383	3473	3371	3178		114	111	109	103
081	Quarrying of stone, sand and clay		343	334	340	331		1579	1498	1387	1302		55	53	48	43
089	Mining and quarrying n.e.c.		485	489	484	466		1804	1976	1984	1876		59	58	61	61
09	Mining support service activities		36	45	48	42		355	368	335	407		15	15	15	15
091	Support activities for petroleum		-	-	-	-		-	-	-	-		-	-	-	-
099	Support activities for other mining, quarrying		36	45	48	42		355	368	335	407		15	15	15	15
B	Mining and quarrying		888	895	897	865		5649	5987	5770	5301		218	229	215	208
35	Electricity, gas, steam & air conditioning		783	816	849	907		13786	13659	13494	13207		665	661	659	668
351	Electric power generation, transmission		368	399	437	477		12411	12160	11935	11595		610	598	593	600
352	Manufacture of gas		11	11	13	13		...	...	45	26		...	...	2	1
353	Steam and air conditioning supply		404	406	399	417		...	...	1514	1587		...	...	64	68
D	Electricity, gas, steam & air conditioning		783	816	849	907		13786	13659	13494	13207		665	661	659	668
36	Water collection, treatment and supply		699	736	717	717		2944	1653	1688	1851		112	67	66	77
37	Sewerage		182	197	199	196		...	...	581	557		...	...	21	21
38	Waste collection,treatment,disposal activities		511	505	501	503		...	...	5839	5713		...	...	203	203
381	Waste collection		251	245	234	226		2304	1833	2323	2158		70	61	71	70
382	Waste treatment and disposal		136	126	127	131		1813	1884	2003	2016		68	72	76	77
383	Materials recovery		124	134	140	146		...	...	1513	1539		...	...	56	56
39	Remediation activities		44	41	43	50		385	...	377	369		13	...	12	13
E	Water supply;sewerage,waste management		1436	1479	1460	1466		9416	7618	8485	8490		334	280	302	314

a/ 071 includes 072.

Finland

ISIC Revision 4		Output (valuation not defined)					Value added at factor values				
			(millions of Euros)					(millions of Euros)			
ISIC	Industry	Note	2012	2013	2014	2015	Note	2012	2013	2014	2015
05	Mining of coal and lignite		-	-	-	-		-	-	-	-
051	Mining of hard coal		-	-	-	-		-	-	-	-
052	Mining of lignite		-	-	-	-		-	-	-	-
06	Extraction of crude petroleum & natural gas		-	-	-	-		-	-	-	-
061	Extraction of crude petroleum		-	-	-	-		-	-	-	-
062	Extraction of natural gas		-	-	-	-		-	-	-	-
07	Mining of metal ores		786	685	499	577		299	170	-149	185
071	Mining of iron ores		-	685a/	499a/	577a/		-	170a/	-149a/	185a/
072	Mining of non-ferrous metal ores		786	...a/	...a/	...a/		299	...a/	...a/	...a/
08	Other mining and quarrying		1047	998	969	911		282	295	303	278
081	Quarrying of stone, sand and clay		424	376	339	308		120	111	103	93
089	Mining and quarrying n.e.c.		623	622	630	603		162	185	200	185
09	Mining support service activities		54	40	40	48		-	2	-3	18
091	Support activities for petroleum		-	-	-	-		-	-	-	-
099	Support activities for other mining, quarrying		54	40	40	48		-	2	-3	18
B	Mining and quarrying		1887	1723	1507	1536		580	467	151	482
35	Electricity, gas, steam & air conditioning		7916	7861	7753	7987		3355	3481	3539	3461
351	Electric power generation, transmission		7304	6793	7055	6926		3108	3193	3219	3137
352	Manufacture of gas		...	...	19	31		...	...	9	8
353	Steam and air conditioning supply		...	...	678	1030		...	...	311	316
D	Electricity, gas, steam & air conditioning		7916	7861	7753	7987		3355	3481	3539	3461
36	Water collection, treatment and supply		915	516	510	634		510	277	284	359
37	Sewerage		...	...	155	148		...	...	72	67
38	Waste collection,treatment,disposal activities		...	...	1623	1625		...	...	465	446
381	Waste collection		332	306	365	340		126	113	133	134
382	Waste treatment and disposal		616	648	641	661		187	198	218	209
383	Materials recovery		...	...	617	624		...	...	114	103
39	Remediation activities		82	...	72	88		30	...	29	30
E	Water supply;sewerage,waste management		2754	2273	2360	2495		1013	758	850	902

a/ 071 includes 072.

France

ISIC Revision 4		Number of enterprises					Number of employees					Wages and salaries paid to employees				
		Note	(number)				Note	(number)				Note	(millions of Euros)			
ISIC	Industry		2012	2013	2014	2015		2012	2013	2014	2015		2012	2013	2014	2015
05	Mining of coal and lignite		-	...	-	1		-	...	-	2		-	...	-	-
051	Mining of hard coal		-	...	-	1		-	...	-	2		-	...	-	-
052	Mining of lignite		-	-	-	-		-	-	-	-		-	-	-	-
06	Extraction of crude petroleum & natural gas		57	58	50	...		692	672	...	...		46	43	16	...
061	Extraction of crude petroleum		50	58a/	45	75		692a/	672a/	...	...		46a/	43a/	16	...
062	Extraction of natural gas		7	...a/	5	...		...a/	...a/	...	...		...a/	...a/	-	...
07	Mining of metal ores		56	...	122	...		...	...	...	...		...	...	18	...
071	Mining of iron ores		56b/	...	122b/	...		...	...	...	...		...	...	18b/	...
072	Mining of non-ferrous metal ores		...b/	84	...b/	...		...	498	...	...		...	...	...b/	...
08	Other mining and quarrying		1543	1637	1624	1822		22780	22329	19812	17809		770	774	691	656
081	Quarrying of stone, sand and clay		1403	1462	1393	1285		21129	20581	18068	16150		701	702	618	572
089	Mining and quarrying n.e.c.		141	175	231	538		1651	1748	1744	1659		70	73	73	84
09	Mining support service activities		62	80	93	98		...	367	472	1198		...	52	169	171
091	Support activities for petroleum		39	45	52	54		...	367c/	374	386		...	52c/	166	150
099	Support activities for other mining, quarrying		23	35	41	44		...	...c/	98	813		...	...c/	3	22
B	Mining and quarrying		1718	1861	1888	2095		24246	23873	21226	19751		877	887	894	861
35	Electricity, gas, steam & air conditioning		18554	20756	24883	27062		175905	177437	183370	180890		8142	8418	8573	8644
351	Electric power generation, transmission		18113	20254	24360	26472		134305	136229	142203	140409		6237	6500	6646	6708
352	Manufacture of gas		93	104	110	158		17225	16871	16696	16132		963	952	962	964
353	Steam and air conditioning supply		347	398	413	432		24376	24337	24471	24349		943	965	966	973
D	Electricity, gas, steam & air conditioning		18554	20756	24883	27062		175905	177437	183370	180890		8142	8418	8573	8644
36	Water collection, treatment and supply		3018	3038	2560	2384		41623	40177	38648	37656		1491	1476	1397	1371
37	Sewerage		1718	1774	1396	1359		15148	15447	14867	13883		473	505	500	483
38	Waste collection,treatment,disposal activities		8322	8686	8622	8118		95063	91664	101884	101093		2978	2898	3282	3283
381	Waste collection		1157	1354	1478	1269		41540	40366	42421	41068		1262	1246	1310	1283
382	Waste treatment and disposal		1078	820	773	861		22246	20343	28990	29590		751	711	1060	1073
383	Materials recovery		6087	6512	6371	5988		31277	30954	30473	30435		966	941	912	927
39	Remediation activities		438	322	277	438		4989	6388	6296	7767		167	208	197	264
E	Water supply;sewerage,waste management		13497	13820	12855	12299		156823	153676	161695	160399		5109	5087	5376	5400

a/ 061 includes 062.
b/ 071 includes 072.
c/ 091 includes 099.

France

ISIC Revision 4		Output (valuation not defined)					Value added at factor values				
			(millions of Euros)					(millions of Euros)			
ISIC	Industry	Note	2012	2013	2014	2015	Note	2012	2013	2014	2015
05	Mining of coal and lignite		-	...	-	-		-	...	-	-
051	Mining of hard coal		-	...	-	-		-	...	-	-
052	Mining of lignite		-	-	-	-		-	-	-	-
06	Extraction of crude petroleum & natural gas		1004	959	446	...		408	280	251	...
061	Extraction of crude petroleum		1004a/	959a/	446	...		408a/	280a/	251	...
062	Extraction of natural gas		...a/	...a/	-	...		...a/	...a/	-	...
07	Mining of metal ores		...	...	96	...		...	...	-29	...
071	Mining of iron ores		...	...	96b/	...		...	...	-29b/	...
072	Mining of non-ferrous metal ores		...	...	...b/	...		...	...	...b/	...
08	Other mining and quarrying		6180	6403	5718	5162		1913	1957	1505	1508
081	Quarrying of stone, sand and clay		5750	5894	5161	4591		1766	1786	1442	1341
089	Mining and quarrying n.e.c.		429	508	557	571		147	172	64	167
09	Mining support service activities		...	121	350	322		...	-79	107	108
091	Support activities for petroleum		...	121c/	331	264		...	-79c/	104	119
099	Support activities for other mining, quarrying		...	...c/	19	58		...	...c/	3	-10
B	Mining and quarrying		7434	7576	6611	5926		2316	2175	1834	1813
35	Electricity, gas, steam & air conditioning		122255	122083	116407	114631		28822	31131	30302	30563
351	Electric power generation, transmission		78706	77991	76877	78605		24272	25986	26233	26191
352	Manufacture of gas		36084	36450	32164	28614		2840	3435	2352	2642
353	Steam and air conditioning supply		7465	7642	7366	7413		1710	1710	1716	1731
D	Electricity, gas, steam & air conditioning		122255	122083	116407	114631		28822	31131	30302	30563
36	Water collection, treatment and supply		12390	12091	11503	11682		3461	3643	3366	3182
37	Sewerage		2525	2804	2728	2768		1204	1331	1243	1212
38	Waste collection,treatment,disposal activities		22995	21080	22938	22527		5762	5303	5710	5633
381	Waste collection		5315	5224	5575	5437		2140	2123	2171	2097
382	Waste treatment and disposal		5485	5261	6659	7135		1758	1560	1938	2041
383	Materials recovery		12196	10594	10703	9955		1865	1620	1602	1495
39	Remediation activities		893	1035	993	1173		286	366	372	463
E	Water supply;sewerage,waste management		38803	37010	38162	38149		10713	10642	10691	10489

a/ 061 includes 062.
b/ 071 includes 072.
c/ 091 includes 099.

Georgia

ISIC Revision 3		Number of establishments					Number of employees					Wages and salaries paid to employees				
			(number)					(number)					(thousands of Georgian Lari)			
ISIC	Industry	Note	2012	2013	2014	2015	Note	2012	2013	2014	2015	Note	2012	2013	2014	2015
101	Mining and agglomeration of hard coal		10	16	20	25		1690	1549	1549	1475		16468	14161	14268	14807
102	Mining and agglomeration of lignite		...	...	...	...		...	...	...	...		...	...	...	...
103	Extraction and agglomeration of peat		1	1	2	1		...	...	...	...		...	...	...	...
111	Extraction of crude oil and natural gas		11	13	15	14		915	969	969	874		15867	17769	17012	18260
112	Oil, gas service activities		7	4	8	5		...	...	...	...		...	...	...	...
120	Mining of uranium and thorium ores		...	...	...	...		...	...	...	...		...	...	...	...
131	Iron ores		...	...	...	...		...	...	...	...		...	...	...	...
132	Non-ferrous metal ores,excl.uranium,thorium		11	16	28	40		1219	1345	1345	2276		19979	21309	20298	40424
141	Quarrying of stone, sand and clay		315	340	408	382		2932	2312	2312	3248		17503	13449	20102	25425
142	Mining and quarrying n.e.c.		8	9	13	12		105	103	...	...		821	654	...	...
C	Total mining and quarrying		363	399	494	479		6918	6289	6289	8018		72597	67400	75014	100775
401	Electricity production, collection, distribution		49	55	57	58		12302	12566	12566	12961		167585	172251	188269	196619
402	Gas production and distribution		55	48	33	25		4330	3967	3967	1605		37406	37572	35671	23320
403	Steam and hot water supply		4	2	3	3		10	24	24	21		68	150	145	158
410	Water collection, purification, distribution		16	17	14	15		5228	5226	5226	5844		38784	46048	52988	58713
E	Total electricity, gas and water supply		124	122	107	101		21870	21783	21783	20431		243844	256021	277072	278811

Georgia

ISIC Revision 3		Output at basic prices					Value added at basic prices				
			(thousands of Georgian Lari)					(thousands of Georgian Lari)			
ISIC	Industry	Note	2012	2013	2014	2015	Note	2012	2013	2014	2015
101	Mining and agglomeration of hard coal		50951	41894	36276	41038		39954	26173	19941	15918
102	Mining and agglomeration of lignite		...	...	...	...		...	...	...	...
103	Extraction and agglomeration of peat		...	...	...	...		...	...	...	...
111	Extraction of crude oil and natural gas		72164	78464	91366	104043		23119	30436	21019	14884
112	Oil, gas service activities		...	...	...	...		...	...	...	...
120	Mining of uranium and thorium ores		...	...	...	...		...	...	...	...
131	Iron ores		...	...	...	...		...	...	...	...
132	Non-ferrous metal ores,excl.uranium,thorium		118107	128984	145085	350496		60738	59406	68878	142835
141	Quarrying of stone, sand and clay		115697	72725	120672	151251		51790	35565	58491	69991
142	Mining and quarrying n.e.c.		3933	3181	...	...		1442	1459	...	...
C	Total mining and quarrying		364390	325332	402249	652729		178643	153104	172659	244938
401	Electricity production, collection, distribution		770814	811341	884557	1037066		413480	501746	537938	548822
402	Gas production and distribution		124388	143782	145779	61158		62023	79782	88343	35972
403	Steam and hot water supply		440	473	520	626		418	382	464	535
410	Water collection, purification, distribution		163062	153637	156743	170396		108139	102625	92164	108189
E	Total electricity, gas and water supply		1058730	1109233	1187599	1269247		584060	684535	718909	693519

Germany

ISIC Revision 4		Number of enterprises					Number of employees					Wages and salaries paid to employees				
			(number)					(thousands)					(millions of Euros)			
ISIC	Industry	Note	2012	2013	2014	2015	Note	2012	2013	2014	2015	Note	2012	2013	2014	2015
051	Mining of hard coal	a/	5	5	5	7	a/	27.0	22.5	20.2	17.5	a/	1100	968	887	817
052	Mining of lignite	a/	...	...	...	...	a/	...	...	...	...	a/	...	...	...	...
061	Extraction of crude petroleum		-	-	-	-		-	-	-	-		-	-	-	-
062	Extraction of natural gas		4	4	4	4		3.8	3.8	4.0	3.9		363	369	376	379
071	Mining of iron ores		-	-	-	-		-	-	-	-		-	-	-	-
072	Mining of non-ferrous metal ores		-	-	-	-		-	-	-	-		-	-	-	-
081	Quarrying of stone, sand and clay		1593	1450	1642	1541		25.4	25.0	27.4	28.3		831	845	961	1028
089	Mining and quarrying n.e.c.		166	155	161	164		5.0	5.0	5.0	4.9		208	211	206	214
091	Support activities for petroleum	b/	133	92	103	97	b/	3.5	3.4	3.2	2.9	b/	167	164	148	158
099	Support activities for other mining, quarrying	b/	...	...	...	...	b/	...	...	...	...	b/	...	...	...	...
B	Mining and quarrying		1902	1706	1916	1813		64.7	59.7	59.7	57.5		2669	2557	2578	2596
351	Electric power generation, transmission		1344	1402	1473	1461		222.3c/	219.9c/	203.1	198.7		10843	10989	11845	11653
352	Manufacture of gas		284	282	281	287		...c/	...c/	...	...		875	781	754	816
353	Steam and air conditioning supply		271	290	304	311		...c/	...c/	...	...		713	718	609	648
D	Electricity, gas, steam & air conditioning		1899	1974	2058	2059		222.3	219.9	228.1	224.6		12432	12487	13208	13116
360	Water collection, treatment and supply		1631	1646	1658	1630		...	...	...	...		1357	1379	1385	1424
370	Sewerage		1404	1415	1418	1419		...	...	...	...		1656	1729	1826	1857
381	Waste collection		765	755	889	861		61.4	63.1	73.0	67.8		1890	1967	2387	2258
382	Waste treatment and disposal		568	559	637	656		37.8	36.9	41.9	48.1		1355	1373	1557	1842
383	Materials recovery		595	605	719	740		24.1	24.3	30.1	31.5		726	749	942	1025
390	Remediation activities		64	66	75	81		...	...	3.6	4.3		99	103	129	160
E	Water supply;sewerage,waste management		5027	5046	5396	5387		204.2	206.3	227.2	230.2		7083	7300	8226	8566

a/ 051 includes 052.
b/ 091 includes 099.
c/ 351 includes 352 and 353.

Germany

ISIC Revision 4		Output (valuation not defined)					Value added at factor values				
			(millions of Euros)					(millions of Euros)			
ISIC	Industry	Note	2012	2013	2014	2015	Note	2012	2013	2014	2015
051	Mining of hard coal	a/	2829	2354	2320	2164	a/	2091	1785	1588	1680
052	Mining of lignite	a/	...	...	...	...	a/	...	...	...	...
061	Extraction of crude petroleum		-	-	-	-		-	-	-	-
062	Extraction of natural gas		2997	2855	2445	2008		1407	1159	901	716
071	Mining of iron ores		-	-	-	-		-	-	-	-
072	Mining of non-ferrous metal ores		-	-	-	-		-	-	-	-
081	Quarrying of stone, sand and clay		4861	4621	5410	5576		1926	1797	2160	2146
089	Mining and quarrying n.e.c.		1211	1360	1070	1213		462	505	389	428
091	Support activities for petroleum	b/	559	496	524	387	b/	272	241	203	159
099	Support activities for other mining, quarrying	b/	...	...	...	...	b/	...	...	...	...
B	Mining and quarrying		12456	11686	11768	11348		6157	5487	5240	5129
351	Electric power generation, transmission		...	...	...	...		39467	37292	34659	32270
352	Manufacture of gas		...	...	...	...		5531	3653	4034	3486
353	Steam and air conditioning supply		...	...	...	...		2549	2358	1775	1833
D	Electricity, gas, steam & air conditioning		...	...	...	...		47547	43302	40468	37589
360	Water collection, treatment and supply		...	...	...	...		4370	4486	4451	4529
370	Sewerage		...	...	...	...		7005	7163	7311	7332
381	Waste collection		...	...	...	...		3946	4015	4694	4383
382	Waste treatment and disposal		...	...	...	...		3412	3439	3817	4405
383	Materials recovery		...	...	...	...		1961	1853	2455	2629
390	Remediation activities		...	...	...	...		157	187	234	320
E	Water supply;sewerage,waste management		...	...	...	...		20851	21143	22962	23596

a/ 051 includes 052.
b/ 091 includes 099.

Greece

ISIC Revision 4		Number of enterprises					Number of employees					Wages and salaries paid to employees				
			(number)					(number)					(millions of Euros)			
ISIC	Industry	Note	2012	2013	2014	2015 a/	Note	2012	2013	2014	2015 a/	Note	2012	2013	2014	2015 a/
051	Mining of hard coal		...	-	-	-		...	-	-	-		...	-	-	-
052	Mining of lignite		...	...	...	12		...	...	...	311		...	...	...	7
061	Extraction of crude petroleum		...	...	...	...		...	...	...	...		...	...	...	...
062	Extraction of natural gas		...	-	-	-		...	-	-	-		...	-	-	-
071	Mining of iron ores		...	-	-	11		...	-	-	2		...	-	-	-
072	Mining of non-ferrous metal ores		...	4	...	19		...	1630	...	1430		...	28	...	46
081	Quarrying of stone, sand and clay		...	353	349	569		...	2852	2999	2454		...	62	65	61
089	Mining and quarrying n.e.c.		...	9	...	...		...	449	...	...		...	12	...	...
091	Support activities for petroleum		...	...	...	9		...	...	...	274		...	...	...	10
099	Support activities for other mining, quarrying		...	...	...	14		...	...	...	52		...	...	...	2
B	Mining and quarrying		380	372	369	682		4467	5417	5742	5192		119	119	136	142
351	Electric power generation, transmission		12	28	24	6767		20317	19820	19512	21329		645	639	772	816
352	Manufacture of gas		...	...	...	14		...	...	...	591		...	...	...	21
353	Steam and air conditioning supply		...	...	...	255		...	...	...	356		...	...	...	6
D	Electricity, gas, steam & air conditioning		18	32	28	7036		21441	20450	20140	22276		687	659	792	842
360	Water collection, treatment and supply		129	128	130	671		8779	8864	9096	8718		237	213	258	227
370	Sewerage		...	...	...	458		...	...	...	809		...	...	...	9
381	Waste collection		...	11	13	532		...	330	281	2151		...	6	5	31
382	Waste treatment and disposal		4	6	...	122		396	426	...	1494		10	9	...	26
383	Materials recovery		12	16	19	239		369	482	538	1771		5	6	7	23
390	Remediation activities		...	...	...	82		...	...	...	160		...	...	...	2
E	Water supply;sewerage,waste management		153	166	173	2104		9722	10202	10490	15103		256	236	281	319

a/ Methodological break in 2015.

Greece

ISIC Revision 4		Output (valuation not defined)					Value added at factor values				
			(millions of Euros)					(millions of Euros)			
ISIC	Industry	Note	2012	2013	2014	2015a/	Note	2012	2013	2014	2015a/
051	Mining of hard coal		...	-	-	-		...	-	-	-
052	Mining of lignite		...	...	...	93		...	...	...	17
061	Extraction of crude petroleum		...	...	...	...		...	...	...	...
062	Extraction of natural gas		...	-	-	-		...	-	-	-
071	Mining of iron ores		...	-	-	-		...	-	-	-
072	Mining of non-ferrous metal ores		...	141	...	306		...	58	...	82
081	Quarrying of stone, sand and clay		...	341	387	384		...	170	195	207
089	Mining and quarrying n.e.c.		...	66	...	...		...	33	...	...
091	Support activities for petroleum		...	...	...	29		...	...	...	15
099	Support activities for other mining, quarrying		...	...	...	5		...	...	...	3
B	Mining and quarrying		622	673	870	915		255	299	335	357
351	Electric power generation, transmission		8711	9643	8303	13464		2155	1998	2316	3084
352	Manufacture of gas		...	...	...	169		...	...	...	145
353	Steam and air conditioning supply		...	...	...	31		...	...	...	13
D	Electricity, gas, steam & air conditioning		9310	9804	8453	13664		2702	2133	2439	3242
360	Water collection, treatment and supply		820	973	788	830		610	791	562	524
370	Sewerage		...	...	...	62		...	...	...	18
381	Waste collection		...	35	26	222		...	14	10	62
382	Waste treatment and disposal		51	...	...	143		20	21	...	55
383	Materials recovery		71	73	69	190		24	17	9	46
390	Remediation activities		...	...	...	10		...	...	...	4
E	Water supply;sewerage,waste management		961	1110	948	1457		665	844	605	708

a/ Methodological break in 2015.

Hungary

ISIC Revision 4		Number of enterprises					Number of employees					Wages and salaries paid to employees				
		Note	(number)				Note	(number)				Note	(millions of Hungarian Forints)			
ISIC	Industry		2012	2013	2014	2015		2012	2013	2014	2015		2012	2013	2014	2015
051	Mining of hard coal		8	7	6	5		16	15	...	1		116	59	31	-
052	Mining of lignite		9	10	8	9		194	223	...	116		492	445	309	279
061	Extraction of crude petroleum		5	4	3	4		1	9	...	10		-	297	185	155
062	Extraction of natural gas		6	6	6	4		78	69	...	57		1012	891	803	620
071	Mining of iron ores		...	...	-	-		...	...	...	-		...	...	-	-
072	Mining of non-ferrous metal ores		...	...	7	6		...	...	...	77		...	...	278	310
081	Quarrying of stone, sand and clay		319	328	318	311		2356	2407	...	2518		6450	6709	7440	7936
089	Mining and quarrying n.e.c.		29	26	26	27		140	124	...	108		202	208	185	186
091	Support activities for petroleum		36	35	32	37		1232	1176	...	1045		7405	7778	8026	7750
099	Support activities for other mining, quarrying		26	23	23	25		111	63	...	121		202	119	123	248
B	Mining and quarrying		448	448	429	428		4226	4170	...	4053		16285	16833	17411	17484
351	Electric power generation, transmission		416	421	412	362		14348	13903	...	13981		99328	99095	103171	106702
352	Manufacture of gas		62	58	67	64		4008	3985	...	3933		22619	22740	24697	24769
353	Steam and air conditioning supply		192	196	189	184		6419	6262	...	6389		24586	25739	25808	25947
D	Electricity, gas, steam & air conditioning		670	675	668	610		24775	24150	...	24303		146534	147574	153676	157418
360	Water collection, treatment and supply		334	306	270	250		20102	21014	...	21981		62941	65222	68781	73439
370	Sewerage		450	410	382	416		3188	2846	...	3415		9892	8965	8273	9021
381	Waste collection		495	458	459	506		7357	5912	...	9101		17557	14131	26827	21979
382	Waste treatment and disposal		224	228	206	209		7485	6483	...	8044		25483	22948	11885	28117
383	Materials recovery		348	336	347	329		2291	2292	...	2652		4773	4958	5372	6262
390	Remediation activities		130	125	136	146		667	614	...	723		2690	2019	2223	2015
E	Water supply;sewerage,waste management		1981	1863	1800	1856		41090	39161	...	45916		123336	118243	123330	140864

Hungary

ISIC Revision 4		Output (valuation not defined)					Value added at factor values				
			(millions of Hungarian Forints)					(millions of Hungarian Forints)			
ISIC	Industry	Note	2012	2013	2014	2015	Note	2012	2013	2014	2015
051	Mining of hard coal		116	119	31	31		-289	-119	...	-
052	Mining of lignite		2285	2791	2408	2945		752	950	...	434
061	Extraction of crude petroleum		810	17783	1976	341		260	4928	...	124
062	Extraction of natural gas		29677	22295	15281	9114		17905	14873	...	3565
071	Mining of iron ores		...	...	-	-		...	...	...	-
072	Mining of non-ferrous metal ores		...	...	1266	1302		...	...	...	682
081	Quarrying of stone, sand and clay		38210	45659	63471	59334		15996	20217	...	27311
089	Mining and quarrying n.e.c.		1070	950	1050	1457		231	327	...	651
091	Support activities for petroleum		32541	32804	32816	38502		17181	17723	...	12121
099	Support activities for other mining, quarrying		868	742	1297	1767		318	327	...	837
B	Mining and quarrying		106907	124121	119594	114793		52962	59671	...	45725
351	Electric power generation, transmission		1151215	1060984	971171	979414		505059	477753	...	472595
352	Manufacture of gas		346174	243790	249036	261981		118506	86597	...	117707
353	Steam and air conditioning supply		254945	242187	230143	227819		26611	16565	...	24025
D	Electricity, gas, steam & air conditioning		1752334	1546960	1450351	1469214		650147	580915	...	614327
360	Water collection, treatment and supply		228594	232924	231255	248527		115584	113404	...	114390
370	Sewerage		66585	60027	53160	58311		32570	29301	...	28334
381	Waste collection		79717	67657	101072	94829		36908	30340	...	38502
382	Waste treatment and disposal		126460	112365	75665	109926		54177	47855	...	57660
383	Materials recovery		45065	50409	56309	60822		11194	11815	...	15996
390	Remediation activities		25946	18287	19541	14787		7087	8134	...	-589
E	Water supply;sewerage,waste management		572368	541669	536970	587202		257519	240851	...	254293

Iceland

ISIC Revision 4		Number of enterprises					Number of employees					Wages and salaries paid to employees				
			(number)					(number)					(millions of Icelandic Kronur)			
ISIC	Industry	Note	2012	2013	2014	2015	Note	2012	2013	2014	2015	Note	2012	2013	2014	2015
051	Mining of hard coal		...	...	...	-		...	...	...	-		...	...	...	-
052	Mining of lignite		...	...	...	-		...	...	...	-		...	...	...	-
061	Extraction of crude petroleum		...	...	...	-		...	...	...	-		...	...	...	-
062	Extraction of natural gas		...	...	...	-		...	...	...	-		...	...	...	-
071	Mining of iron ores		...	...	...	-		...	...	...	-		...	...	...	-
072	Mining of non-ferrous metal ores		...	...	...	-		...	...	...	-		...	...	...	-
081	Quarrying of stone, sand and clay		...	...	...	24		...	...	...	94		...	...	...	644
089	Mining and quarrying n.e.c.		...	...	...	8		...	...	...	45		...	...	...	234
091	Support activities for petroleum		...	...	...	1		...	...	...	1		...	...	...	-
099	Support activities for other mining, quarrying		...	...	...	-		...	...	...	-		...	...	...	-
B	Mining and quarrying		...	...	...	33		...	...	...	140		...	...	...	878
351	Electric power generation, transmission		...	...	...	31		...	...	...	1189		...	...	...	10226
352	Manufacture of gas		...	...	...	-		...	...	...	-		...	...	...	-
353	Steam and air conditioning supply		...	...	...	61		...	...	...	382		...	...	...	3277
D	Electricity, gas, steam & air conditioning		...	...	...	92		...	...	...	1571		...	...	...	13503
360	Water collection, treatment and supply		...	...	...	10		...	...	...	2		...	...	...	-
370	Sewerage		...	...	...	3		...	...	...	12		...	...	...	44
381	Waste collection		...	...	...	22		...	...	...	508		...	...	...	2780
382	Waste treatment and disposal		...	...	...	8		...	...	...	64		...	...	...	351
383	Materials recovery		...	...	...	16		...	...	...	124		...	...	...	658
390	Remediation activities		...	...	...	-		...	...	...	-		...	...	...	-
E	Water supply;sewerage,waste management		...	...	...	59		...	...	...	709		...	...	...	3848

Iceland

ISIC Revision 4		Output (valuation not defined)					Value added at factor values				
			(millions of Icelandic Kronur)					(millions of Icelandic Kronur)			
ISIC	Industry	Note	2012	2013	2014	2015	Note	2012	2013	2014	2015
051	Mining of hard coal		...	...	...	-		...	...	...	-
052	Mining of lignite		...	...	...	-		...	...	...	-
061	Extraction of crude petroleum		...	...	...	-		...	...	...	-
062	Extraction of natural gas		...	...	...	-		...	...	...	-
071	Mining of iron ores		...	...	...	-		...	...	...	-
072	Mining of non-ferrous metal ores		...	...	...	-		...	...	...	-
081	Quarrying of stone, sand and clay		...	...	...	2867		...	...	...	1419
089	Mining and quarrying n.e.c.		...	...	...	1229		...	...	...	497
091	Support activities for petroleum		...	...	...	-		...	...	...	-
099	Support activities for other mining, quarrying		...	...	...	-		...	...	...	-
B	Mining and quarrying		...	...	...	4096		...	...	...	1917
351	Electric power generation, transmission		...	...	...	77232		...	...	...	61607
352	Manufacture of gas		...	...	...	-		...	...	...	-
353	Steam and air conditioning supply		...	...	...	38345		...	...	...	25690
D	Electricity, gas, steam & air conditioning		...	...	...	115577		...	...	...	87297
360	Water collection, treatment and supply		...	...	...	88		...	...	...	29
370	Sewerage		...	...	...	146		...	...	...	88
381	Waste collection		...	...	...	8676		...	...	...	4433
382	Waste treatment and disposal		...	...	...	1463		...	...	...	424
383	Materials recovery		...	...	...	2385		...	...	...	732
390	Remediation activities		...	...	...	-		...	...	...	-
E	Water supply;sewerage,waste management		...	...	...	12757		...	...	...	5691

India

ISIC Revision 4		Output at basic prices					Value added at basic prices				
			(billions of Indian Rupees)					(billions of Indian Rupees)			
ISIC	Industry	Note	2012	2013	2014	2015	Note	2012	2013	2014	2015
051	Mining of hard coal	a/	988.2	996.4	1059.8	1102.9		...	...	...	...
052	Mining of lignite	a/	...	...	...	...		...	...	...	...
061	Extraction of crude petroleum	b/	2383.0	2665.2	2436.1	1944.5		...	...	...	...
062	Extraction of natural gas	b/	...	...	...	...		...	...	...	...
071	Mining of iron ores		492.8	733.7	779.7	625.2		...	...	...	...
072	Mining of non-ferrous metal ores		161.4	249.0	288.7	320.9		...	...	...	...
081	Quarrying of stone, sand and clay	c/	895.7	917.5	1041.8	1125.6		...	...	...	...
089	Mining and quarrying n.e.c.	c/	...	...	...	...		...	...	...	...
091	Support activities for petroleum		...	...	...	...		...	...	...	...
099	Support activities for other mining, quarrying		...	...	...	...		...	...	...	...
B	Mining and quarrying		4921.1	5561.8	5606.2	5119.1		2858.4	2957.9	3141.8	2962.5
351	Electric power generation, transmission		4781.9	5480.7	6001.5	6740.7		1682.0	2048.5	2176.8	2518.5
352	Manufacture of gas		802.3	977.9	1019.5	764.3		156.5	168.7	153.2	145.5
353	Steam and air conditioning supply		...	...	...	...		...	...	...	...
D	Electricity, gas, steam & air conditioning	d/	6258.1	7301.9	8068.1	8760.8	d/	2153.5	2601.6	2796.3	3216.5
360	Water collection, treatment and supply		313.1	424.1	557.9	731.9		175.1	220.6	281.3	354.3
370	Sewerage		...	...	...	...		...	...	...	...
381	Waste collection		...	...	...	...		...	...	...	...
382	Waste treatment and disposal		...	...	...	...		...	...	...	...
383	Materials recovery		...	...	...	...		...	...	...	...
390	Remediation activities		360.9	419.3	489.2	524.0		139.9	163.7	185.0	198.3
E	Water supply;sewerage,waste management	d/	...	...	...	...	d/	...	...	...	...

a/ 051 includes 052.
b/ 061 includes 062.
c/ 081 includes 089.
d/ D includes E.

Iran (Islamic Republic of)

ISIC Revision 4		Number of establishments					Number of employees					Wages and salaries paid to employees				
			(number)					(number)					(billions of Iranian Rials)			
ISIC	Industry	Note	2012	2013	2014	2015	Note	2012	2013	2014	2015	Note	2012	2013	2014	2015
05	Mining of coal and lignite		96	101	100	91		12102	12257	11205	9948		1421	1583	1712	1611
051	Mining of hard coal		96	101	100	91		12102	12257	11205	9948		1421	1583	1712	1611
052	Mining of lignite		-	-	...	...		...	...	...	...		...	-	...	...
06	Extraction of crude petroleum & natural gas		...	...	...	...		...	...	...	...		...	...	...	...
061	Extraction of crude petroleum		...	...	...	...		...	...	...	...		...	...	...	...
062	Extraction of natural gas		...	...	...	...		...	...	...	...		...	...	...	...
07	Mining of metal ores		243	306	313	281		27352	32895	34077	37400		3820	5801	8783	...
071	Mining of iron ores		122	159	145	131		12102	18396	18356	20839		2198	3109	4836	...
072	Mining of non-ferrous metal ores		121	147	168	150		15250	14499	15688	16561		1622	2692	3948	5034
08	Other mining and quarrying		4977	5038	4942	4842		42204	45085	42327	39933		3616	4679	5302	5779
081	Quarrying of stone, sand and clay		4443	4463	4365	4317		38039	40148	37614	35363		3293	4165	4742	5174
089	Mining and quarrying n.e.c.		533	575	577	525		4165	4937	4713	4570		323	514	559	605
09	Mining support service activities		...	...	...	...		...	...	...	...		...	...	...	...
091	Support activities for petroleum		...	...	...	...		...	...	...	...		...	...	...	...
099	Support activities for other mining, quarrying		...	...	...	...		...	...	...	...		...	...	...	...
B	Mining and quarrying		5316	5445	5355	5214		81176	90236	87576	87280		8858	12064	15797	20723
35	Electricity, gas, steam & air conditioning		...	...	...	...		...	...	...	...		...	...	...	...
351	Electric power generation, transmission		...	...	...	...		...	...	...	...		...	...	...	...
352	Manufacture of gas		...	...	...	...		...	...	...	...		...	...	...	...
353	Steam and air conditioning supply		...	...	...	...		...	...	...	...		...	...	...	...
D	Electricity, gas, steam & air conditioning		...	...	...	...		...	...	...	...		...	...	...	...
36	Water collection, treatment and supply		...	...	...	...		...	...	...	...		...	...	...	...
37	Sewerage		...	...	...	...		...	...	...	...		...	...	...	...
38	Waste collection,treatment,disposal activities		...	...	...	...		...	...	...	...		...	...	...	...
381	Waste collection		...	...	...	...		...	...	...	...		...	...	...	...
382	Waste treatment and disposal		...	...	...	...		...	...	...	...		...	...	...	...
383	Materials recovery		...	...	...	...		...	...	...	...		...	...	...	...
39	Remediation activities		...	...	...	...		...	...	...	...		...	...	...	...
E	Water supply;sewerage,waste management		...	...	...	...		...	...	...	...		...	...	...	...

Iran (Islamic Republic of)

ISIC Revision 4		Output at producers' prices					Value added at producers' prices				
			(billions of Iranian Rials)					(billions of Iranian Rials)			
ISIC	Industry	Note	2012	2013	2014	2015	Note	2012	2013	2014	2015
05	Mining of coal and lignite		3997.4	4327.2	4768.4	5613.4		3317.7	2688.5	3255.7	4917.3
051	Mining of hard coal		3997.4	4327.2	4768.4	5613.4		3317.7	2688.5	3255.7	4917.3
052	Mining of lignite		...	-	...	...		...	...	...	...
06	Extraction of crude petroleum & natural gas		...	...	...	...		...	...	...	...
061	Extraction of crude petroleum		...	...	...	...		...	...	...	...
062	Extraction of natural gas		...	...	...	...		...	...	...	...
07	Mining of metal ores		78810.8	99606.9	109215.8	85991.4		61979.4	72525.5	74104.8	58510.6
071	Mining of iron ores		51695.5	74049.7	78636.5	52785.4		38740.7	52610.9	52697.5	34569.4
072	Mining of non-ferrous metal ores		27115.3	25557.2	30579.2	33206.0		23238.7	19914.6	21407.3	23941.2
08	Other mining and quarrying		19573.3	25408.6	28637.4	28795.4		...	18218.1	20229.9	20643.6
081	Quarrying of stone, sand and clay		17453.2	22624.0	25155.1	25181.4		...	16166.7	17616.8	17673.4
089	Mining and quarrying n.e.c.		2120.1	2784.5	3482.3	3614.0		...	2051.4	2613.1	2970.1
09	Mining support service activities		...	...	...	...		...	...	...	...
091	Support activities for petroleum		...	...	...	...		...	...	...	...
099	Support activities for other mining, quarrying		...	...	...	...		...	...	...	...
B	Mining and quarrying		102381.4	129342.6	142621.5	120400.2		79163.5	93432.2	97590.4	84071.4
35	Electricity, gas, steam & air conditioning		...	...	...	...		...	...	...	...
351	Electric power generation, transmission		...	...	...	...		...	...	...	...
352	Manufacture of gas		...	...	...	...		...	...	...	...
353	Steam and air conditioning supply		...	...	...	...		...	...	...	...
D	Electricity, gas, steam & air conditioning		...	...	...	...		...	...	...	...
36	Water collection, treatment and supply		...	...	...	...		...	...	...	...
37	Sewerage		...	...	...	...		...	...	...	...
38	Waste collection,treatment,disposal activities		...	...	...	...		...	...	...	...
381	Waste collection		...	...	...	...		...	...	...	...
382	Waste treatment and disposal		...	...	...	...		...	...	...	...
383	Materials recovery		...	...	...	...		...	...	...	...
39	Remediation activities		...	...	...	...		...	...	...	...
E	Water supply;sewerage,waste management		...	...	...	...		...	...	...	...

Ireland

ISIC Revision 4		Number of enterprises					Number of employees					Wages and salaries paid to employees				
			(number)					(number)					(millions of Euros)			
ISIC	Industry	Note	2012	2013	2014	2015	Note	2012	2013	2014	2015	Note	2012	2013	2014	2015
051	Mining of hard coal		...	...	...	...		...	...	...	...		...	...	...	...
052	Mining of lignite		-	-	-	-		-	-	-	-		-	-	-	-
061	Extraction of crude petroleum		13	14	15	...		8	3	9	...		1	-	1	...
062	Extraction of natural gas		...	...	...	...		...	...	...	...		...	...	...	...
071	Mining of iron ores		-	-	-	-		-	-	-	-		-	-	-	-
072	Mining of non-ferrous metal ores		14	...	...	...		1280	...	...	...		97	...	...	...
081	Quarrying of stone, sand and clay		322a/	325a/	315a/	247		2661a/	2684a/	2699a/	1448		121a/	133a/	135a/	83
089	Mining and quarrying n.e.c.		...a/	...a/	...a/	...		...a/	...a/	...a/	...		...a/	...a/	...a/	...
091	Support activities for petroleum		35	41	39	35		24	51	25	20		2	3	2	1
099	Support activities for other mining, quarrying		21	19	21	21		47	35	46	28		3	2	3	2
B	Mining and quarrying		411	421	411	405		4076	4048	3965	4069		226	232	229	241
351	Electric power generation, transmission		...	...	...	...		...	...	...	...		...	...	...	...
352	Manufacture of gas		...	...	...	...		...	...	...	...		...	...	...	...
353	Steam and air conditioning supply		9	9	9	10		...	...	...	...		1	1	1	1
D	Electricity, gas, steam & air conditioning		456	485	483	515		8693	8203	8456	8807		602	554	599	652
360	Water collection, treatment and supply		158	168	173	...		206	237	552	...		8	8	21	...
370	Sewerage		189	188	192	...		721	752	786	...		28	28	31	...
381	Waste collection		242	234	224	224		2385	2527	2694	2794		87	83	91	132
382	Waste treatment and disposal		113	108	104	104		2165	2105	2057	2141		75	70	74	82
383	Materials recovery		256	259	251	254		1684	1509	1591	1678		56	48	53	76
390	Remediation activities		27	28	31	35		125	87	80	121		5	3	3	6
E	Water supply;sewerage,waste management		985	985	975	1014		7286	7217	7760	8474		258	241	273	385

a/ 081 includes 089.

Ireland

ISIC Revision 4		Output (valuation not defined)					Value added at factor values				
			(millions of Euros)					(millions of Euros)			
ISIC	Industry	Note	2012	2013	2014	2015	Note	2012	2013	2014	2015
051	Mining of hard coal		...	...	...	...		...	...	...	...
052	Mining of lignite		-	-	-	-		-	-	-	-
061	Extraction of crude petroleum		9	2	3	...		3	1	2	...
062	Extraction of natural gas		...	...	...	...		...	...	...	...
071	Mining of iron ores		-	-	-	-		-	-	-	-
072	Mining of non-ferrous metal ores		651	...	...	...		311	...	...	...
081	Quarrying of stone, sand and clay		441a/	445a/	459a/	304		181a/	224a/	239a/	100
089	Mining and quarrying n.e.c.		...a/	...a/	...a/	...		...a/	...a/	...a/	...
091	Support activities for petroleum		99	35	7	10		-7	11	4	4
099	Support activities for other mining, quarrying		16	12	10	5		6	3	4	2
B	Mining and quarrying		1304	1070	1007	878		535	521	495	269
351	Electric power generation, transmission		...	...	...	...		...	...	...	...
352	Manufacture of gas		...	...	...	...		...	...	...	...
353	Steam and air conditioning supply		2	2	2	2		1	1	1	1
D	Electricity, gas, steam & air conditioning		7254	7373	6990	6922		3546	3492	3540	3287
360	Water collection, treatment and supply		19	108	57	...		9	-61	22	...
370	Sewerage		145	150	146	...		42	53	57	...
381	Waste collection		321	332	466	424		126	124	156	148
382	Waste treatment and disposal		275	344	361	390		104	128	114	159
383	Materials recovery		314	308	309	302		92	69	85	99
390	Remediation activities		21	13	17	20		8	5	5	8
E	Water supply;sewerage,waste management		1095	1254	1355	1390		381	317	440	414b/

a/ 081 includes 089.
b/ Sum of available data.

Israel

ISIC Revision 4		Number of enterprises a/					Number of employees a/					Wages and salaries paid to employees a/				
			(number)					(thousands)					(millions of Israeli New Sheqalim)			
ISIC	Industry	Note	2012	2013	2014	2015	Note	2012	2013	2014	2015	Note	2012	2013	2014	2015
051	Mining of hard coal		...	...	...	...		...	...	...	...		...	...	...	...
052	Mining of lignite		...	...	...	...		...	...	...	...		...	...	...	...
061	Extraction of crude petroleum		...	...	...	...		...	...	...	...		...	...	...	...
062	Extraction of natural gas		...	...	...	...		...	...	...	...		...	...	...	...
071	Mining of iron ores		...	...	...	...		...	...	...	...		...	...	...	...
072	Mining of non-ferrous metal ores		...	...	...	...		...	...	...	...		...	...	...	...
081	Quarrying of stone, sand and clay		93	90	79	82		1	1	1	1		150	158	152	168
089	Mining and quarrying n.e.c.		8	7	7	9		2	2	2	2		852	747	795	622
091	Support activities for petroleum		60	63	62	57		-	-	-	-		...	81	61	70
099	Support activities for other mining, quarrying		11	27	40	66		-	-	-	-		2	3	4	7
B	Mining and quarrying		176	191	193	219		4	4	4	4		...	1055	1096	950
351	Electric power generation, transmission		314	373	419	466		14	14	14	14		4058	4101	4408	4351
352	Manufacture of gas		...	...	...	...		...	...	...	...		...	...	...	...
353	Steam and air conditioning supply		...	...	...	...		...	...	...	...		...	...	...	...
D	Electricity, gas, steam & air conditioning		314	373	419	466		14	14	14	14		4058	4101	4408	4351
360	Water collection, treatment and supply		267	277	289	295		4	4	4	4		747	794	817	829
370	Sewerage		147	146	127	138		1	1	1	1		147	144	164	186
381	Waste collection		198	208	234	243		4	5	5	5		392	406	461	470
382	Waste treatment and disposal		56	49	50	51		1	1	1	1		91	121	129	132
383	Materials recovery		111	133	160	162		1	1	1	1		90	106	117	129
390	Remediation activities		-	4	7	9		-	-	-	-		...	...	7	7
E	Water supply;sewerage,waste management		779	817	867	898		11	12	13	13		1467	1571	1695	1753

a/ Data were derived from OECD's databases.

Israel

ISIC Revision 4		Output (valuation not defined) a/					Value added (valuation not defined) a/				
		Note	(millions of Israeli New Sheqalim)				Note	(millions of Israeli New Sheqalim)			
ISIC	Industry		2012	2013	2014	2015		2012	2013	2014	2015
051	Mining of hard coal		...	...	...	...		...	...	...	...
052	Mining of lignite		...	...	...	...		...	...	...	...
061	Extraction of crude petroleum		...	...	...	...		...	...	...	...
062	Extraction of natural gas		...	...	...	...		...	...	...	...
071	Mining of iron ores		...	...	...	...		...	...	...	...
072	Mining of non-ferrous metal ores		...	...	...	...		...	...	...	...
081	Quarrying of stone, sand and clay		1736	2042	2288	...		620	800	947	...
089	Mining and quarrying n.e.c.		...	...	...	...		...	...	...	...
091	Support activities for petroleum		...	...	...	...		...	...	...	...
099	Support activities for other mining, quarrying		...	...	...	...		...	...	...	...
B	Mining and quarrying		14210	14172	13724	...		7497	8569	8930	...
351	Electric power generation, transmission		...	...	...	...		...	...	...	...
352	Manufacture of gas		...	...	...	...		...	...	...	...
353	Steam and air conditioning supply		...	...	...	...		...	...	...	...
D	Electricity, gas, steam & air conditioning		...	...	...	...		...	...	...	...
360	Water collection, treatment and supply		...	...	...	...		...	...	...	...
370	Sewerage		...	...	...	...		...	...	...	...
381	Waste collection		...	...	...	...		...	...	...	...
382	Waste treatment and disposal		...	...	...	...		...	...	...	...
383	Materials recovery		...	...	...	...		...	...	...	...
390	Remediation activities		...	...	...	...		...	...	...	...
E	Water supply;sewerage,waste management		42131	42338	43334	...		10120	19200	21588	...

a/ Data were derived from OECD's databases.

Italy

ISIC Revision 4		Number of enterprises					Number of employees					Wages and salaries paid to employees				
			(number)					(thousands)					(millions of Euros)			
ISIC	Industry	Note	2012	2013	2014	2015	Note	2012	2013	2014	2015	Note	2012	2013	2014	2015
051	Mining of hard coal		...	...	...	-		...	...	...	-		...	...	...	-
052	Mining of lignite		-	-	-	-		-	-	-	-		-	-	-	-
061	Extraction of crude petroleum		4	5	6	5		0.3	0.4	0.4	0.6		24	29	33	42
062	Extraction of natural gas		8	7	7	7		12.6	12.0	12.4	12.1		722	793	831	839
071	Mining of iron ores		-	-	-	-		-	-	-	-		-	-	-	-
072	Mining of non-ferrous metal ores		...	...	...	4		...	...	...	-		...	...	...	1
081	Quarrying of stone, sand and clay		2119	2025	2032	1968		12.6	12.0	12.1	11.8		323	312	325	335
089	Mining and quarrying n.e.c.		268	249	168	150		2.7	2.5	1.9	1.7		81	73	73	56
091	Support activities for petroleum		44	42	39	52		1.7	1.8	1.9	2.2		94	100	106	123
099	Support activities for other mining, quarrying		-	-	-	-		-	-	-	-		-	-	-	-
B	Mining and quarrying		2451	2336	2257	2186		30.4	29.1	29.3	28.4		1262	1322	1382	1394
351	Electric power generation, transmission		8129	9336	9622	9925		58.5	57.6	58.8	60.0		2715	2680	2708	2628
352	Manufacture of gas		593	618	622	620		22.0	22.0	21.4	20.6		877	887	870	845
353	Steam and air conditioning supply		204	215	215	230		3.7	3.7	3.6	3.6		136	133	137	138
D	Electricity, gas, steam & air conditioning		8926	10169	10459	10775		84.3	83.3	83.9	84.2		3729	3700	3715	3612
360	Water collection, treatment and supply		858	873	844	829		28.6	30.0	29.2	30.6		989	1043	1044	1119
370	Sewerage		1395	1428	1409	1402		8.9	8.4	7.8	8.1		271	256	240	253
381	Waste collection		2126	2123	2129	2100		91.1	89.4	90.9	90.7		2568	2581	2610	2617
382	Waste treatment and disposal		856	903	917	942		18.3	17.9	18.8	20.5		605	586	624	681
383	Materials recovery		3157	3199	3146	3163		22.0	22.4	22.4	22.4		554	572	571	586
390	Remediation activities		575	595	701	795		5.6	6.2	6.0	6.6		178	212	204	226
E	Water supply;sewerage,waste management		8967	9121	9146	9231		174.6	174.4	175.0	179.0		5164	5250	5292	5482

Italy

ISIC Revision 4		Output (valuation not defined)					Value added at factor values				
		Note	(millions of Euros)				Note	(millions of Euros)			
ISIC	Industry		2012	2013	2014	2015		2012	2013	2014	2015
051	Mining of hard coal		...	...	...	-		...	...	...	-
052	Mining of lignite		-	-	-	-		-	-	-	-
061	Extraction of crude petroleum		2164	1996	1891	1333		1510	1246	1063	568
062	Extraction of natural gas		66117	42607	40222	33201		2058	1556	1429	1622
071	Mining of iron ores		-	-	-	-		-	-	-	-
072	Mining of non-ferrous metal ores		...	...	...	2		...	...	...	-1
081	Quarrying of stone, sand and clay		2368	2196	2356	2346		740	724	768	833
089	Mining and quarrying n.e.c.		691	614	581	543		182	167	169	147
091	Support activities for petroleum		493	488	1674	2750		203	212	260	422
099	Support activities for other mining, quarrying		-	-	-	-		-	-	-	-
B	Mining and quarrying		71870	47950	46765	40176		4707	3941	3716	3591
351	Electric power generation, transmission		88181	80780	79104	84048		23737	21502	20850	20042
352	Manufacture of gas		41633	14659	17760	15035		5372	4769	4554	4399
353	Steam and air conditioning supply		2062	2198	1797	1672		448	615	415	436
D	Electricity, gas, steam & air conditioning		131877	97637	98661	100754		29557	26886	25818	24877
360	Water collection, treatment and supply		7209	7543	7384	8166		2969	3305	3446	3932
370	Sewerage		1756	1585	1521	1580		644	605	576	614
381	Waste collection		10634	10224	10293	10639		4580	4553	4862	4734
382	Waste treatment and disposal		4902	5231	5439	5610		1802	1933	2040	2199
383	Materials recovery		8300	8091	7844	7303		1488	1406	1384	1464
390	Remediation activities		1091	1425	1271	1403		241	179	68	259
E	Water supply;sewerage,waste management		33891	34100	33751	34701		11724a/	11981	12375	13203

a/ Sum of available data.

Japan

ISIC Revision 4		Establishments					Number of employees					Wages and salaries paid to employees				
		Note	(number)				Note	(number)				Note	(millions of Japanese Yen)			
ISIC	Industry		2012	2013	2014	2015		2012	2013	2014	2015		2012	2013	2014	2015
051	Mining of hard coal		...	...	...	...		...	...	...	...		...	...	...	...
052	Mining of lignite		...	...	...	...		...	...	...	...		...	...	...	...
061	Extraction of crude petroleum		...	...	...	...		...	...	...	...		...	...	...	...
062	Extraction of natural gas		...	...	...	...		...	...	...	...		...	...	...	...
071	Mining of iron ores		...	...	...	...		...	...	...	...		...	...	...	...
072	Mining of non-ferrous metal ores		...	...	...	...		...	...	...	...		...	...	...	...
081	Quarrying of stone, sand and clay		...	...	...	...		1635a/	1614a/	1724a/	...		4466a/	4390a/	4717a/	...
089	Mining and quarrying n.e.c.		...	...	...	...		...a/	...a/	...a/	...		...a/	...a/	...a/	...
091	Support activities for petroleum		...	...	...	...		...	...	...	...		...	...	...	...
099	Support activities for other mining, quarrying		...	...	...	...		...	...	...	...		...	...	...	...
B	Mining and quarrying		...	...	...	...		1635	1614	1724	...		4466	4390	4717	...
351	Electric power generation, transmission		...	...	...	...		1842b/	1850b/	1995b/	...		1370b/	1679b/	2168b/	...
352	Manufacture of gas		...	...	...	...		...b/	...b/	...b/	...		...b/	...b/	...b/	...
353	Steam and air conditioning supply		...	...	...	...		...b/	...b/	...b/	...		...b/	...b/	...b/	...
D	Electricity, gas, steam & air conditioning		...	...	...	...		1842	1850	1995	...		1370	1679	2168	...
360	Water collection, treatment and supply		...	...	...	...		...	...	...	...		...	...	...	...
370	Sewerage		...	...	...	...		...	...	...	...		...	...	...	...
381	Waste collection		...	...	...	...		13494c/	13257c/	13716c/	...		23897c/	22488c/	24212c/	...
382	Waste treatment and disposal		...	...	...	...		...c/	...c/	...c/	...		...c/	...c/	...c/	...
383	Materials recovery		...	...	...	...		...c/	...c/	...c/	...		...c/	...c/	...c/	...
390	Remediation activities		...	...	...	...		...	...	...	...		...	...	...	...
E	Water supply;sewerage,waste management		...	...	...	...		13494	13257	13716	...		23897	22488	24212	...

a/ 081 includes 089.
b/ 351 includes 352 and 353.
c/ 381 includes 382 and 383.

Japan

ISIC Revision 4		Output (valuation not defined)					Value added (valuation not defined)				
		Note	(millions of Japanese Yen)				Note	(millions of Japanese Yen)			
ISIC	Industry		2012	2013	2014	2015		2012	2013	2014	2015
051	Mining of hard coal		...	...	...	...		...	...	...	...
052	Mining of lignite		...	...	...	...		...	...	...	...
061	Extraction of crude petroleum		...	...	...	...		...	...	...	...
062	Extraction of natural gas		...	...	...	...		...	...	...	...
071	Mining of iron ores		...	...	...	...		...	...	...	...
072	Mining of non-ferrous metal ores		...	...	...	...		...	...	...	...
081	Quarrying of stone, sand and clay		42273a/	42611a/	42849a/	...		20500a/	21925a/	21356a/	...
089	Mining and quarrying n.e.c.		...a/	...a/	...a/	...		...a/	...a/	...a/	...
091	Support activities for petroleum		...	...	...	...		...	...	...	...
099	Support activities for other mining, quarrying		...	...	...	...		...	...	...	...
B	Mining and quarrying		42273	42611	42849	...		20500	21925	21356	...
351	Electric power generation, transmission		24551b/	27207b/	42719b/	...		17685b/	19268b/	28664b/	...
352	Manufacture of gas		...b/	...b/	...b/	...		...b/	...b/	...b/	...
353	Steam and air conditioning supply		...b/	...b/	...b/	...		...b/	...b/	...b/	...
D	Electricity, gas, steam & air conditioning		24551	27207	42719	...		17685	19268	28664	...
360	Water collection, treatment and supply		...	...	...	...		...	...	...	...
370	Sewerage		...	...	...	...		...	...	...	...
381	Waste collection		792942c/	832318c/	875554c/	...		190981c/	177524c/	190722c/	...
382	Waste treatment and disposal		...c/	...c/	...c/	...		...c/	...c/	...c/	...
383	Materials recovery		...c/	...c/	...c/	...		...c/	...c/	...c/	...
390	Remediation activities		...	...	...	...		...	...	...	...
E	Water supply;sewerage,waste management		792942	832318	875554	...		190981	177524	190722	...

a/ 081 includes 089.
b/ 351 includes 352 and 353.
c/ 381 includes 382 and 383.

Jordan

ISIC Revision 4		Number of establishments					Number of persons engaged					Wages and salaries paid to employees				
			(number)					(number)					(thousands of Jordanian Dinars)			
ISIC	Industry	Note	2011	2012	2013	2014	Note	2011	2012	2013	2014	Note	2011	2012	2013	2014
05	Mining of coal and lignite		...	...	...	...		...	...	...	...		...	...	...	...
051	Mining of hard coal		...	...	...	...		...	...	...	...		...	...	...	...
052	Mining of lignite		...	...	...	...		...	...	...	...		...	...	...	...
06	Extraction of crude petroleum & natural gas		1	...	...	1		143	...	...	180		2092	...	...	2320
061	Extraction of crude petroleum		1a/	...	...	...		143a/	...	...	...		2092a/	...	...	...
062	Extraction of natural gas		...a/	...	...	1		...a/	...	...	180		...a/	...	...	2320
07	Mining of metal ores		...	...	...	...		...	...	...	...		...	...	...	...
071	Mining of iron ores		...	...	...	...		...	...	...	...		...	...	...	...
072	Mining of non-ferrous metal ores		...	...	...	...		...	...	...	...		...	...	...	...
08	Other mining and quarrying		82	...	...	112		5319	...	...	5909		97598	97598	...	82407
081	Quarrying of stone, sand and clay		82b/	...	...	107		5319b/	...	...	725		97598b/	97598b/	...	2360
089	Mining and quarrying n.e.c.		...b/	...	...	5		...b/	...	...	5184		...b/	...b/	...	80047
09	Mining support service activities		...	...	...	...		...	...	...	...		...	...	...	...
091	Support activities for petroleum		...	...	...	...		...	...	...	...		...	...	...	...
099	Support activities for other mining, quarrying		...	...	...	...		...	...	...	...		...	...	...	...
B	Mining and quarrying		83	...	...	113		5462	...	...	6089		99690	97598	...	84727
35	Electricity, gas, steam & air conditioning		9	...	...	9		7415	...	...	7822		79370	...	...	71272
351	Electric power generation, transmission		9c/	...	...	9		7415c/	...	...	7822		79370c/	...	...	71272
352	Manufacture of gas		...c/	...	...	...		...c/	...	...	...		...c/	...	...	...
353	Steam and air conditioning supply		...c/	...	...	...		...c/	...	...	...		...c/	...	...	...
D'	Electricity, gas, steam & air conditioning	d/	9	...	...	9	d/	7415	...	...	7822	d/	79370	...	...	71272
36	Water collection, treatment and supply		6	...	...	13		12	...	...	40		20	...	...	54
37	Sewerage		...	...	...	...		...	...	...	...		...	...	...	...
38	Waste collection,treatment,disposal activities		39	...	...	42		176	...	...	205		431	...	...	451
381	Waste collection		39e/	...	...	...		176e/	...	...	...		431e/	...	...	...
382	Waste treatment and disposal		...e/	...	...	...		...e/	...	...	...		...e/	...	...	...
383	Materials recovery		...e/	...	...	42		...e/	...	...	205		...e/	...	...	451
39	Remediation activities		...	...	...	...		...	...	...	...		...	...	...	...
E	Water supply;sewerage,waste management	d/	45	...	...	55	d/	188	...	...	245	d/	451	...	...	505

a/ 061 includes 062.
b/ 081 includes 089.
c/ 351 includes 352 and 353.
d/ Sum of available data.
e/ 381 includes 382 and 383.

Jordan

ISIC Revision 4		Output (valuation not defined)					Value added (valuation not defined)				
		Note	(thousands of Jordanian Dinars)				Note	(thousands of Jordanian Dinars)			
ISIC	Industry		2011	2012	2013	2014		2011	2012	2013	2014
05	Mining of coal and lignite		...	...	...	...		...	...	...	...
051	Mining of hard coal		...	...	...	...		...	...	...	...
052	Mining of lignite		...	...	...	...		...	...	...	...
06	Extraction of crude petroleum & natural gas		9323	...	...	6733		8848	...	...	6474
061	Extraction of crude petroleum		9323a/	...	...	...		8848a/	...	...	...
062	Extraction of natural gas		...a/	...	...	6733		...a/	...	...	6474
07	Mining of metal ores		...	...	...	...		...	...	...	...
071	Mining of iron ores		...	...	...	...		...	...	...	...
072	Mining of non-ferrous metal ores		...	...	...	...		...	...	...	...
08	Other mining and quarrying		1349651	1198495	...	950259		1022703	892411	...	726587
081	Quarrying of stone, sand and clay		1349651b/	1198495b/	...	11411		1022703b/	892411b/	...	4399
089	Mining and quarrying n.e.c.		...b/	...b/	...	938849		...b/	...b/	...	722187
09	Mining support service activities		...	...	...	...		...	...	...	...
091	Support activities for petroleum		...	...	...	...		...	...	...	...
099	Support activities for other mining, quarrying		...	...	...	...		...	...	...	...
B	Mining and quarrying		1358974	1198495	...	956993		1031551	892411	...	733061
35	Electricity, gas, steam & air conditioning		186028	...	...	278832		...	...	...	...
351	Electric power generation, transmission		186028c/	...	...	278832		...	...	...	...
352	Manufacture of gas		...c/	...	...	...		...	...	...	...
353	Steam and air conditioning supply		...c/	...	...	...		...	...	...	...
D	Electricity, gas, steam & air conditioning	d/	186028	...	...	278832		...	...	...	...
36	Water collection, treatment and supply		204	...	...	399		161	...	...	138
37	Sewerage		...	...	...	...		...	...	...	...
38	Waste collection,treatment,disposal activities		4484	...	...	2791		1056	...	...	980
381	Waste collection		4484e/	...	...	...		1056e/	...	...	...
382	Waste treatment and disposal		...e/	...	...	...		...e/	...	...	...
383	Materials recovery		...e/	...	...	2791		...e/	...	...	980
39	Remediation activities		...	...	...	...		...	...	...	...
E	Water supply;sewerage,waste management	d/	4688	...	...	3190	d/	1217	...	...	1118

a/ 061 includes 062.
b/ 081 includes 089.
c/ 351 includes 352 and 353.
d/ Sum of available data.
e/ 381 includes 382 and 383.

Kazakhstan

ISIC Revision 4		Number of establishments					Number of employees					Wages and salaries paid to employees				
			(number)					(number)					(millions of Kazakhstan Tenge)			
ISIC	Industry	Note	2012	2013	2014	2015	Note	2012	2013	2014	2015	Note	2012	2013	2014	2015
051	Mining of hard coal		81	87	87	88		31726	31599	30840	29781		51760.0	53371.0	57553.0	60174.0
052	Mining of lignite		12	12	13	15		2309	2315	2154	1902		2256.0	2608.0	2420.0	2442.0
061	Extraction of crude petroleum		250	272	287	314		44349	45373	46340	46695		164414.0	213490.0	272321.0	304898.0
062	Extraction of natural gas		12	13	18	19		...	122	162	192		129.0	...	364.0	453.0
071	Mining of iron ores		44	44	46	51		21674	22083	21667	21109		31074.0	35773.0	40052.0	38693.0
072	Mining of non-ferrous metal ores		260	309	349	374		63847	62303	60424	58819		103904.0	114694.0	123794.0	131801.0
081	Quarrying of stone, sand and clay		1067	1177	1266	1348		9999	9440	7789	6264		9804.0	9692.0	8858.0	7200.0
089	Mining and quarrying n.e.c.		167	178	194	203		6960	6994	7074	7542		6191.0	6704.0	7653.0	9759.0
091	Support activities for petroleum		523	580	626	640		32826	35296	35128	30808		87371.0	107563.0	126293.0	116676.0
099	Support activities for other mining, quarrying		74	112	155	190		2864	3232	2556	1657		5842.0	6906.0	5703.0	3020.0
B	Mining and quarrying		2490	2784	3041	3242		216554	218757	214134	204769		463096.0	552329.0	646734.0	677271.0
351	Electric power generation, transmission		779	805	828	853		80389	80672	79743	81953		93675.0	106503.0	116580.0	128458.0
352	Manufacture of gas		159	163	156	170		8132	7527	7108	4578		11276.0	12610.0	15235.0	10024.0
353	Steam and air conditioning supply		296	300	288	296		27301	27681	30106	28453		22611.0	25968.0	32020.0	33259.0
D	Electricity, gas, steam & air conditioning		1234	1268	1272	1319		115822	115880	116957	114984		127525.0	145048.0	163833.0	171449.0
360	Water collection, treatment and supply		628	655	646	657		25145	27634	28857	27647		17948.0	22519.0	26964.0	28029.0
370	Sewerage		60	65	79	91		2701	2165	2413	2285		1982.0	1875.0	2221.0	2247.0
381	Waste collection		626	641	653	663		6602	7378	8058	6925		4397.0	5327.0	6752.0	6365.0
382	Waste treatment and disposal		216	244	294	318		2488	2246	2349	1886		2792.0	2902.0	2967.0	2742.0
383	Materials recovery		464	488	494	491		1822	1796	1604	857		1646.0	1589.0	1594.0	1005.0
390	Remediation activities		71	84	98	107		214	262	281	80		268.0	317.0	447.0	192.0
E	Water supply;sewerage,waste management		2065	2177	2264	2327		38972	41481	43562	39680		29014.0	34517.0	40962.0	40577.0

Kazakhstan

ISIC Revision 4		Output at basic prices					Value added at basic prices				
			(millions of Kazakhstan Tenge)					(millions of Kazakhstan Tenge)			
ISIC	Industry	Note	2012	2013	2014	2015	Note	2012	2013	2014	2015
051	Mining of hard coal	a/	331353	335970	356450	358712	a/	136697	143962	154058	155283
052	Mining of lignite	a/	...	...	...	...	a/	...	...	...	...
061	Extraction of crude petroleum	b/	8789779	9103777	9248728	5981451	b/	4240751	4352073	4707646	3917034
062	Extraction of natural gas	b/	...	...	...	...	b/	...	...	...	...
071	Mining of iron ores	c/	1467804	1692050	1788119	1692601	c/	607384	661839	720665	684082
072	Mining of non-ferrous metal ores	c/	...	...	...	...	c/	...	...	...	...
081	Quarrying of stone, sand and clay	d/	148650	131963	150937	169018	d/	84164	84749	99887	116410
089	Mining and quarrying n.e.c.	d/	...	...	...	...	d/	...	...	...	...
091	Support activities for petroleum	e/	480589	553785	675665	531314	e/	219745	235071	300516	297759
099	Support activities for other mining, quarrying	e/	...	...	...	...	e/	...	...	...	...
B	Mining and quarrying		11218175	11817544	12219898	8733096		5288741	5477694	5982772	5170568
351	Electric power generation, transmission		...	...	...	...		...	...	...	...
352	Manufacture of gas		...	...	...	...		...	...	...	...
353	Steam and air conditioning supply		...	...	...	52198		...	...	...	9205
D	Electricity, gas, steam & air conditioning		1108819	1225279	1339036	1403477		518617	580318	637529	693589
360	Water collection, treatment and supply		...	...	...	...		...	...	...	...
370	Sewerage		...	...	...	...		...	...	...	...
381	Waste collection		...	...	...	...		...	...	...	...
382	Waste treatment and disposal		...	...	...	...		...	...	...	...
383	Materials recovery		...	...	...	...		...	...	...	...
390	Remediation activities		...	...	...	...		...	...	...	...
E	Water supply;sewerage,waste management		230634	220121	228696	224749		99608	99498	103972	101807

a/ 051 includes 052.
b/ 061 includes 062.
c/ 071 includes 072.
d/ 081 includes 089.
e/ 091 includes 099.

Kuwait

ISIC Revision 3		Number of establishments					Number of employees					Wages and salaries paid to employees				
			(number)					(number)					(millions of Kuwaiti Dinars)			
ISIC	Industry	Note	2011	2012	2013	2014	Note	2011	2012	2013	2014	Note	2011	2012	2013	2014
101	Mining and agglomeration of hard coal		...	...	...	...		...	...	...	...		...	...	...	...
102	Mining and agglomeration of lignite		...	...	...	...		...	...	...	...		...	...	...	...
103	Extraction and agglomeration of peat		...	...	...	...		...	...	...	...		...	...	...	...
111	Extraction of crude oil and natural gas		3	3	3	...		10853	11191	12221	...		533.9	614.2	618.5	...
112	Oil, gas service activities		4	4	4	...		4207	4600	4812	...		34.5	35.9	39.3	...
120	Mining of uranium and thorium ores		...	...	...	...		...	...	...	...		...	...	...	...
131	Iron ores		...	...	...	...		...	...	...	...		...	...	...	...
132	Non-ferrous metal ores,excl.uranium,thorium		...	...	...	...		...	...	...	...		...	...	...	...
141	Quarrying of stone, sand and clay		...	...	...	...		...	...	...	...		...	...	...	...
142	Mining and quarrying n.e.c.		...	...	...	...		...	...	...	...		...	...	...	...
C	Total mining and quarrying		7a/	7a/	7a/	7		15060a/	15791a/	17033a/	18881		568.4a/	650.1a/	657.7a/	830.0
401	Electricity production, collection, distribution		...	...	...	...		...	...	...	...		...	...	...	...
402	Gas production and distribution		...	...	...	...		...	...	...	...		...	...	...	...
403	Steam and hot water supply		...	...	...	...		...	...	...	...		...	...	...	...
410	Water collection, purification, distribution		...	...	...	...		...	...	...	...		...	...	...	...
E	Total electricity, gas and water supply		...	...	...	...		...	...	...	...		...	...	...	...

a/ Sum of available data.

Kuwait

ISIC Revision 3		Output at producers' prices					Value added at producers' prices				
		Note	(millions of Kuwaiti Dinars)				Note	(millions of Kuwaiti Dinars)			
ISIC	Industry		2011	2012	2013	2014		2011	2012	2013	2014
101	Mining and agglomeration of hard coal		...	...	...	...		...	...	...	...
102	Mining and agglomeration of lignite		...	...	...	...		...	...	...	...
103	Extraction and agglomeration of peat		...	...	...	...		...	...	...	...
111	Extraction of crude oil and natural gas		27615.6	32697.7	32253.3	...		26896.1	31764.8	31297.7	...
112	Oil, gas service activities		170.7	163.8	169.9	...		88.7	84.1	85.9	...
120	Mining of uranium and thorium ores		...	...	...	...		...	...	...	...
131	Iron ores		...	...	...	...		...	...	...	...
132	Non-ferrous metal ores,excl.uranium,thorium		...	...	...	...		...	...	...	...
141	Quarrying of stone, sand and clay		...	...	...	...		...	...	...	...
142	Mining and quarrying n.e.c.		...	...	...	...		...	...	...	...
C	Total mining and quarrying		27786.3a/	32861.5a/	32423.2a/	28906.0		26984.8a/	31848.9a/	31383.6a/	28157.0
401	Electricity production, collection, distribution		...	...	...	...		...	...	...	...
402	Gas production and distribution		...	...	...	...		...	...	...	...
403	Steam and hot water supply		...	...	...	...		...	...	...	...
410	Water collection, purification, distribution		...	...	...	...		...	...	...	...
E	Total electricity, gas and water supply		...	...	...	...		...	...	...	...

a/ Sum of available data.

Latvia

ISIC Revision 4		Number of establishments					Number of employees					Wages and salaries paid to employees				
			(number)					(number)					(thousands of Euros)			
ISIC	Industry	Note	2012	2013	2014	2015	Note	2012	2013	2014	2015	Note	2012	2013	2014	2015
051	Mining of hard coal		...	...	...	...		...	...	...	...		...	...	...	...
052	Mining of lignite		...	...	...	...		...	...	...	...		...	...	...	...
061	Extraction of crude petroleum		1	1	1	...		16	15	19	...		...	...	...	...
062	Extraction of natural gas		...	...	...	...		...	...	...	...		...	...	...	...
071	Mining of iron ores		...	...	...	...		...	...	...	...		...	...	...	...
072	Mining of non-ferrous metal ores		...	...	...	...		...	...	...	...		...	...	...	...
081	Quarrying of stone, sand and clay		162	168	157	...		784	895	949	...		3395	4290	...	...
089	Mining and quarrying n.e.c.		59	59	72	...		2197	2109	2147	...		13089	14816	21986	...
091	Support activities for petroleum		1	1	1	...		4	1	1	...		...	...	...	...
099	Support activities for other mining, quarrying		7	7	5	...		47	58	51	...		...	...	...	...
B	Mining and quarrying		230	236	236	...		3048	3078	3167	...		16731	19364	28502	...
351	Electric power generation, transmission		328	346	371	...		5287	5409	5234	...		52652	54912	75431	...
352	Manufacture of gas		6	10	10	...		1328	1313	1325	...		...	...	...	...
353	Steam and air conditioning supply		119	124	125	...		4129	4383	4012	...		...	...	...	...
D	Electricity, gas, steam & air conditioning		454	480	506	...		10744	11105	10571	...		95200	99186	141130	...
360	Water collection, treatment and supply		63	72	73	...		1701	1786	1758	...		10798	11808	17097	...
370	Sewerage		64	68	71	...		1473	1571	1643	...		7011	7827	12876	...
381	Waste collection		96	104	107	...		2894	3021	3163	...		13457	14033	20990	...
382	Waste treatment and disposal		30	30	27	...		362	458	404	...		2470	3014	4319	...
383	Materials recovery		73	78	71	...		605	624	548	...		2662	2361	3266	...
390	Remediation activities		15	19	19	...		81	86	98	...		228	257	400	...
E	Water supply;sewerage,waste management		342	372	368	...		7115	7547	7614	...		36626	39300	58948	...

Latvia

ISIC Revision 4		Output at producers' prices					Value added at factor values				
			(thousands of Euros)					(thousands of Euros)			
ISIC	Industry	Note	2012	2013	2014	2015	Note	2012	2013	2014	2015
051	Mining of hard coal		...	...	...	...		...	...	...	...
052	Mining of lignite		...	...	...	...		...	...	...	...
061	Extraction of crude petroleum		...	...	...	...		...	...	...	...
062	Extraction of natural gas		...	...	...	...		...	...	...	...
071	Mining of iron ores		...	...	...	...		...	...	...	...
072	Mining of non-ferrous metal ores		...	...	...	...		...	...	...	...
081	Quarrying of stone, sand and clay		44370	46565	...	...		15601	16215	...	...
089	Mining and quarrying n.e.c.		89035	100492	147956	...		38658	39490	56627	...
091	Support activities for petroleum		...	...	...	...		...	...	...	...
099	Support activities for other mining, quarrying		...	...	...	...		...	...	...	...
B	Mining and quarrying		134689	148696	218984	...		54954	56552	77169	...
351	Electric power generation, transmission		862587	883009	1085264	...		279270	277952	348435	...
352	Manufacture of gas		...	...	...	...		...	...	...	...
353	Steam and air conditioning supply		...	...	...	...		...	...	...	...
D	Electricity, gas, steam & air conditioning		1178811	1225442	1511944	...		430393	459150	573480	...
360	Water collection, treatment and supply		48367	47715	72081	...		28126	30109	46981	...
370	Sewerage		32400	34585	51428	...		19209	21531	33975	...
381	Waste collection		71779	71870	96570	...		25947	31657	44377	...
382	Waste treatment and disposal		17438	21441	26911	...		8746	10122	14447	...
383	Materials recovery		66991	39918	53435	...		6471	4711	9391	...
390	Remediation activities		1800	1800	2314	...		772	702	854	...
E	Water supply;sewerage,waste management		238775	217329	302739	...		89271	98832	150026	...

Liechtenstein

ISIC Revision 4		Number of establishments					Number of employees					Wages and salaries			
			(number)					(number)					(Swiss Francs)		
ISIC	Industry	Note	2012	2013	2014	2015	Note	2012	2013	2014	2015	Note	2012	2013	2014	2015
05	Mining of coal and lignite		...	...	...	...		...	...	...	...		...	...	...	...
06	Extraction of crude petroleum & natural gas		...	...	...	...		...	...	...	...		...	...	...	...
07	Mining of metal ores		...	...	...	...		...	...	...	...		...	...	...	...
08	Other mining and quarrying		...	...	...	...		...	...	...	...		...	...	...	...
09	Mining support service activities		...	...	...	...		...	...	...	...		...	...	...	...
B	Mining and quarrying		5	5	5	5		53	52	53	50		...	...	...	...
35	Electricity, gas, steam & air conditioning		...	...	...	...		...	...	...	...		...	...	...	...
D	Electricity, gas, steam & air conditioning		3	4	4	3		233	228	218	201		...	...	...	...
36	Water collection, treatment and supply		...	...	...	...		...	...	...	...		...	...	...	...
37	Sewerage		...	...	...	...		...	...	...	...		...	...	...	...
38	Waste collection,treatment,disposal activities		...	...	...	...		...	...	...	...		...	...	...	...
39	Remediation activities		...	...	...	...		...	...	...	...		...	...	...	...
E	Water supply;sewerage,waste management		11	11	13	13		123	117	110	105		...	...	...	...

Lithuania

ISIC Revision 4		Number of establishments					Number of employees					Wages and salaries paid to employees				
			(number)					(number)					(thousands of Euros)			
ISIC	Industry	Note	2012	2013	2014	2015	Note	2012	2013	2014	2015	Note	2012	2013	2014	2015
051	Mining of hard coal		-	-	-	-		-	-	-	-		-	-	-	-
052	Mining of lignite		-	-	-	-		-	-	-	-		-	-	-	-
061	Extraction of crude petroleum		4	4	4	4		279	269	255	212		4906	4903	4576	3536
062	Extraction of natural gas		-	-	-	-		-	-	-	-		-	-	-	-
071	Mining of iron ores		-	-	-	-		-	-	-	-		-	-	-	-
072	Mining of non-ferrous metal ores		-	-	-	-		-	-	-	-		-	-	-	-
081	Quarrying of stone, sand and clay		61	68	86	86		1158	1229	1311	1358		9425	11013	12407	13657
089	Mining and quarrying n.e.c.		25	25	26	26		1090	1090	1153	1148		7897	8757	9670	10603
091	Support activities for petroleum		-	-	-	-		-	-	-	-		-	-	-	-
099	Support activities for other mining, quarrying		-	-	-	-		-	-	-	-		-	-	-	-
B	Mining and quarrying		90	97	116	116		2527	2588	2719	2718		22229	24674	26652	27796
351	Electric power generation, transmission		400	1115	1327	1385		4670	5260	5621	5757		52197	54441	54500	56745
352	Manufacture of gas		8	14	15	14		1749	1980	1799	1603		23167	23990	24765	24310
353	Steam and air conditioning supply		80	85	92	90		6227	6240	6093	6063		54217	57380	57017	59919
D	Electricity, gas, steam & air conditioning		488	1214	1434	1489		12646	13480	13513	13423		129580	135810	136282	140975
360	Water collection, treatment and supply		76	75	72	73		5654	5620	5624	5624		41544	43183	45457	47536
370	Sewerage		44	47	49	49		570	665	704	620		3364	4717	4781	4387
381	Waste collection		115	117	121	122		3414	3459	3464	3319		18076	20003	20602	21722
382	Waste treatment and disposal		77	75	86	92		1574	1709	1838	2079		10665	12040	13401	15758
383	Materials recovery		83	78	82	85		1581	1543	1588	1473		10682	10753	11799	10915
390	Remediation activities		-	-	-	-		-	-	-	-		-	-	-	-
E	Water supply;sewerage,waste management		395	392	410	421		12793	12996	13218	13115		84330	90696	96039	100319

Lithuania

ISIC Revision 4		Output at producers' prices					Value added at factor values				
			(thousands of Euros)					(thousands of Euros)			
ISIC	Industry	Note	2012	2013	2014	2015	Note	2012	2013	2014	2015
051	Mining of hard coal		-	-	-	-		-	-	-	-
052	Mining of lignite		-	-	-	-		-	-	-	-
061	Extraction of crude petroleum		76026	63689	56898	31450		43554	33203	27540	12825
062	Extraction of natural gas		-	-	-	-		-	-	-	-
071	Mining of iron ores		-	-	-	-		-	-	-	-
072	Mining of non-ferrous metal ores		-	-	-	-		-	-	-	-
081	Quarrying of stone, sand and clay		66899	92546	98781	84414		28253	39024	37029	37333
089	Mining and quarrying n.e.c.		52704	60745	60024	65230		19218	22015	24746	25100
091	Support activities for petroleum		-	-	-	-		-	-	-	-
099	Support activities for other mining, quarrying		-	-	-	-		-	-	-	-
B	Mining and quarrying		195629	216980	215703	181095		91026	94242	89315	75258
351	Electric power generation, transmission		1129558	1269094	1168839	1126000		327250	378224	395130	380598
352	Manufacture of gas		130516	143090	252826	161184		95996	83664	191715	66304
353	Steam and air conditioning supply		660879	626587	561876	528461		170779	167302	163190	177720
D	Electricity, gas, steam & air conditioning		1920953	2038770	1983541	1815645		594025	629189	750035	624622
360	Water collection, treatment and supply		148829	157571	172826	175081		106415	116492	138907	119139
370	Sewerage		13301	29574	28064	25718		6206	10298	10447	8824
381	Waste collection		88755	95772	94765	104316		33829	36154	38613	39522
382	Waste treatment and disposal		70927	80089	104221	119418		29264	32961	39152	46156
383	Materials recovery		155018	133906	168505	127613		30728	26771	33408	23971
390	Remediation activities		-	-	-	-		-	-	-	-
E	Water supply;sewerage,waste management		476829	496911	568380	552146		206443	222676	260527	237612

Luxembourg

ISIC Revision 4		Number of enterprises					Number of employees					Wages and salaries paid to employees				
		Note	(number)				Note	(number)				Note	(millions of Euros)			
ISIC	Industry		2012	2013	2014	2015		2012	2013	2014	2015		2012	2013	2014	2015
051	Mining of hard coal		-	-	-	-		-	-	-	-		-	-	-	-
052	Mining of lignite		-	-	-	-		-	-	-	-		-	-	-	-
061	Extraction of crude petroleum		-	-	-	-		-	-	-	-		-	-	-	-
062	Extraction of natural gas		-	-	-	-		-	-	-	-		-	-	-	-
071	Mining of iron ores		-	-	-	-		-	-	-	-		-	-	-	-
072	Mining of non-ferrous metal ores		-	-	-	-		-	-	-	-		-	-	-	-
081	Quarrying of stone, sand and clay		10	10	10	10		275	266	252	248		12	12	12	12
089	Mining and quarrying n.e.c.		-	-	-	-		-	-	-	-		-	-	-	-
091	Support activities for petroleum		-	-	-	-		-	-	-	-		-	-	-	-
099	Support activities for other mining, quarrying		-	-	-	-		-	-	-	-		-	-	-	-
B	Mining and quarrying		10	10	10	10		275	266	252	248		12	12	12	12
351	Electric power generation, transmission		56	55	62	62		1341	1435	1470a/	1529a/		100	107	...	121a/
352	Manufacture of gas		4	5	5	5		...	...	...a/	...a/		...	...	...	...a/
353	Steam and air conditioning supply		14	14	15	13		155	168	...a/	...a/		7	8	...	...a/
D	Electricity, gas, steam & air conditioning		74	74	82	80		1341	1435	1470	1529		100	107	116	121
360	Water collection, treatment and supply		8	8	8	7		140	139	138	144		9	10	10	11
370	Sewerage		6	6	6	4		...	...	...	...		...	...	...	...
381	Waste collection		11	14	14	13		591	601	639	632		23	24	26	26
382	Waste treatment and disposal		13	13	15	16		127	129	147	...		7	7	7	...
383	Materials recovery		22	22	22	20		298	295	281	...		12	13	12	...
390	Remediation activities		4	4	4	4		...	...	...	...		...	...	...	...
E	Water supply;sewerage,waste management		64	67	69	64		1197	1210	1260	1243		52	55	57	58

a/ 351 includes 352 and 353.

Luxembourg

ISIC Revision 4		Output (valuation not defined)					Value added at factor values				
			(millions of Euros)					(millions of Euros)			
ISIC	Industry	Note	2012	2013	2014a/	2015a/	Note	2012	2013	2014	2015
051	Mining of hard coal		-	-	-	-		-	-	-	-
052	Mining of lignite		-	-	-	-		-	-	-	-
061	Extraction of crude petroleum		-	-	-	-		-	-	-	-
062	Extraction of natural gas		-	-	-	-		-	-	-	-
071	Mining of iron ores		-	-	-	-		-	-	-	-
072	Mining of non-ferrous metal ores		-	-	-	-		-	-	-	-
081	Quarrying of stone, sand and clay		75	75	71	74		32	32	30	27
089	Mining and quarrying n.e.c.		-	-	-	-		-	-	-	-
091	Support activities for petroleum		-	-	-	-		-	-	-	-
099	Support activities for other mining, quarrying		-	-	-	-		-	-	-	-
B	Mining and quarrying		75	75	71	74		32	32	30	27
351	Electric power generation, transmission		950	959	...	1129b/		225	276	434b/	509b/
352	Manufacture of gas		...	...	...	...b/		...	...	...b/	...b/
353	Steam and air conditioning supply		82	89	...	...b/		28	29	...b/	...b/
D	Electricity, gas, steam & air conditioning		950	959	1055	1129		225	276	434	509
360	Water collection, treatment and supply		42	43	43	47		23	24	23	23
370	Sewerage		...	...	...	...		...	...	...	...
381	Waste collection		85	95	98	104		39	45	49	47
382	Waste treatment and disposal		21	21	32	...		11	10	16	...
383	Materials recovery		49	47	45	...		24	22	22	...
390	Remediation activities		...	...	...	...		...	...	...	...
E	Water supply;sewerage,waste management		205	215	226	254		101	103	111	124

a/ Provisional data.
b/ 351 includes 352 and 353.

Malaysia

ISIC Revision 4		Number of establishments					Number of employees					Wages and salaries paid to employees				
			(number)					(number)					(millions of Malaysian Ringgits)			
ISIC	Industry	Note	2012	2013	2014	2015	Note	2012	2013	2014	2015	Note	2012	2013	2014	2015
05	Mining of coal and lignite		...	...	...	8		...	...	...	1316		...	...	...	42.0
051	Mining of hard coal		...	...	...	8		...	...	...	1316		...	...	...	42.0
052	Mining of lignite		...	...	...	...		...	...	...	...		...	...	...	...
06	Extraction of crude petroleum & natural gas		16	...	...	26		15015	16355	17350	17950		3691.8	4157.0	4931.0	5108.0
061	Extraction of crude petroleum		10	...	...	16		12409	...	...	14395		3244.2	...	...	4442.0
062	Extraction of natural gas		6	...	...	10		2606	...	...	3555		447.6	...	...	666.0
07	Mining of metal ores		...	...	...	150		...	...	...	11172		...	...	...	351.0
071	Mining of iron ores		...	...	...	35		...	...	...	1714		...	...	...	44.0
072	Mining of non-ferrous metal ores		...	...	...	115		...	...	...	9458		...	...	...	306.0
08	Other mining and quarrying		...	...	...	672		...	...	...	26925		...	...	...	782.0
081	Quarrying of stone, sand and clay		...	...	...	622		...	...	...	25399		...	...	...	739.0
089	Mining and quarrying n.e.c.		...	...	...	50		...	...	...	1526		...	...	...	44.0
09	Mining support service activities		...	...	...	170		...	...	...	24991		...	...	...	1587.0
091	Support activities for petroleum		...	...	...	131		...	...	...	23748		...	...	...	1552.0
099	Support activities for other mining, quarrying		...	...	...	39		...	...	...	1243		...	...	...	34.0
B	Mining and quarrying		16a/	...	...	1026		15015a/	16355a/	17350a/	82354		3691.8a/	4157.0a/	4931.0a/	7870.0
35	Electricity, gas, steam & air conditioning		...	...	...	305		41975	...	42875	47907		2450.0	...	2789.0	3622.0
351	Electric power generation, transmission		...	...	...	122		...	...	...	39878		2299.0	...	2610.0	3380.0
352	Manufacture of gas		...	...	...	17		...	...	...	1163		40.0	...	54.0	77.0
353	Steam and air conditioning supply		...	...	...	166		...	...	...	6866		111.0	...	125.0	165.0
D	Electricity, gas, steam & air conditioning		...	...	...	305		41975a/	...	42875a/	47907		2450.0	...	2789.0	3622.0
36	Water collection, treatment and supply		...	...	...	...		...	...	...	...		...	...	...	...
37	Sewerage		...	...	...	...		...	...	...	...		...	...	...	...
38	Waste collection,treatment,disposal activities		...	...	...	...		...	...	...	...		...	...	...	...
381	Waste collection		...	...	...	...		...	...	...	...		...	...	...	...
382	Waste treatment and disposal		...	...	...	...		...	...	...	...		...	...	...	...
383	Materials recovery		...	...	...	...		...	...	...	...		...	...	...	...
39	Remediation activities		...	...	...	...		...	...	...	...		...	...	...	...
E	Water supply;sewerage,waste management		...	...	...	...		...	...	...	...		...	...	...	...

a/ Sum of available data.

Malaysia

ISIC Revision 4		Output at basic prices					Value added at basic prices				
		Note	(millions of Malaysian Ringgits)				Note	(millions of Malaysian Ringgits)			
ISIC	Industry		2012	2013	2014	2015		2012	2013	2014	2015
05	Mining of coal and lignite		...	...	...	429.0		...	...	...	88.0
051	Mining of hard coal		...	...	...	429.0		...	...	...	88.0
052	Mining of lignite		...	...	...	...		...	...	...	...
06	Extraction of crude petroleum & natural gas		109369.8	115096.0	119054.0	108893.0		94441.6	98308.0	103568.0	92361.0
061	Extraction of crude petroleum		81221.6	...	...	79425.0		69366.2	...	...	66950.0
062	Extraction of natural gas		28148.2	...	...	29467.0		25075.5	...	...	25411.0
07	Mining of metal ores		...	...	...	3118.0		...	...	...	1373.0
071	Mining of iron ores		...	...	...	366.0		...	...	...	133.0
072	Mining of non-ferrous metal ores		...	...	...	2753.0		...	...	...	1240.0
08	Other mining and quarrying		...	...	...	6893.0		...	...	...	2657.0
081	Quarrying of stone, sand and clay		...	...	...	6590.0		...	...	...	2512.0
089	Mining and quarrying n.e.c.		...	...	...	303.0		...	...	...	145.0
09	Mining support service activities		...	...	...	11734.0		...	...	...	6580.0
091	Support activities for petroleum		...	...	...	11507.0		...	...	...	6472.0
099	Support activities for other mining, quarrying		...	...	...	227.0		...	...	...	109.0
B	Mining and quarrying		109369.8a/	115096.0a/	119054.0a/	131067.0		94441.6a/	98308.0a/	103568.0a/	103059.0
35	Electricity, gas, steam & air conditioning		38945.0	...	53104.0	54389.0		21132.0	...	25598.0	26710.0
351	Electric power generation, transmission		36248.0	...	49725.0	50460.0		20280.0	...	23845.0	24637.0
352	Manufacture of gas		2109.0	...	2576.0	2838.0		552.0	...	1378.0	1581.0
353	Steam and air conditioning supply		588.0	...	803.0	1092.0		299.0	...	375.0	493.0
D	Electricity, gas, steam & air conditioning		38945.0	...	53104.0	54389.0		21132.0	...	25598.0	26710.0
36	Water collection, treatment and supply		...	...	...	...		...	...	...	...
37	Sewerage		...	...	...	...		...	...	...	...
38	Waste collection,treatment,disposal activities		...	...	...	...		...	...	...	...
381	Waste collection		...	...	...	...		...	...	...	...
382	Waste treatment and disposal		...	...	...	...		...	...	...	...
383	Materials recovery		...	...	...	...		...	...	...	...
39	Remediation activities		...	...	...	...		...	...	...	...
E	Water supply;sewerage,waste management		...	...	...	...		...	...	...	...

a/ Sum of available data.

Malta

ISIC Revision 4		Number of enterprises					Number of employees					Wages and salaries paid to employees				
			(number)					(number)					(thousands of Euros)			
ISIC	Industry	Note	2011	2012	2013	2014	Note	2011	2012	2013	2014	Note	2011	2012	2013	2014
05	Mining of coal and lignite		-	-	-	-		-	-	-	-		-	-	-	-
06	Extraction of crude petroleum & natural gas		-	-	-	-		-	-	-	-		-	-	-	-
07	Mining of metal ores		-	-	-	-		-	-	-	-		-	-	-	-
08	Other mining and quarrying		46	42	40	...		194	...	...	157		2341	...	...	2555
09	Mining support service activities		...	...	...	...		...	...	...	...		...	...	...	...
B	Mining and quarrying		...	...	...	...		...	...	192	...		...	...	...	...
35	Electricity, gas, steam & air conditioning		...	...	...	...		...	...	...	...		...	...	...	...
D	Electricity, gas, steam & air conditioning		...	...	...	...		...	...	...	...		...	...	...	...
36	Water collection, treatment and supply		...	...	...	...		...	...	...	...		...	...	...	...
37	Sewerage		...	...	...	...		...	...	...	4		...	...	...	49
38	Waste collection,treatment,disposal activities		...	...	110	...		480	...	458	378		5000	...	5187	4925
39	Remediation activities		...	...	...	...		...	...	...	...		...	...	...	...
E	Water supply;sewerage,waste management		160	164	153	159		...	...	...	...		...	...	...	...

Malta

ISIC Revision 4		Output at factor values					Value added at factor values				
		Note	(thousands of Euros)				Note	(thousands of Euros)			
ISIC	Industry		2011	2012	2013	2014		2011	2012	2013	2014
05	Mining of coal and lignite		-	-	-	-		-	-	-	-
06	Extraction of crude petroleum & natural gas		-	-	-	-		-	-	-	-
07	Mining of metal ores		-	-	-	-		-	-	-	-
08	Other mining and quarrying		12419	...	...	11179		4049	...	...	4170
09	Mining support service activities		...	...	...	...		...	...	...	...
B	Mining and quarrying		...	...	...	...		...	...	...	...
35	Electricity, gas, steam & air conditioning		...	...	...	...		...	...	...	...
D	Electricity, gas, steam & air conditioning		...	...	...	...		...	...	...	...
36	Water collection, treatment and supply		...	...	...	...		...	...	...	...
37	Sewerage		...	392	473	806		...	291	257	420
38	Waste collection,treatment,disposal activities		42129	37397	36804	41407		14831	11996	10131	12728
39	Remediation activities		...	...	...	...		...	...	...	...
E	Water supply;sewerage,waste management		...	...	...	...		...	...	...	...

Mauritius

ISIC Revision 4		Number of establishments					Number of employees					Wages and salaries paid to employees				
			(number)					(number)					(millions of Mauritian Rupees)			
ISIC	Industry	Note	2012	2013	2014	2015	Note	2012	2013	2014	2015	Note	2012	2013	2014	2015
05	Mining of coal and lignite		...	...	...	...		...	...	...	...		...	...	...	...
06	Extraction of crude petroleum & natural gas		...	...	...	...		...	...	...	...		...	...	...	...
07	Mining of metal ores		...	...	...	...		...	...	...	...		...	...	...	...
08	Other mining and quarrying		...	...	...	...		...	...	...	...		...	...	...	...
09	Mining support service activities		...	...	...	...		...	...	...	...		...	...	...	...
B	Mining and quarrying		25	24	24	23		975	1020	1046	1035		473	455	472	423
35	Electricity, gas, steam & air conditioning		7	7	7	7		2373	2384	2323	2275		1238	1460	1561	1626
D	Electricity, gas, steam & air conditioning		7	7	7	7		2373	2384	2323	2275		1238	1460	1561	1626
36	Water collection, treatment and supply		...	...	...	...		...	...	...	...		...	...	...	...
37	Sewerage		...	...	...	...		...	...	...	...		...	...	...	...
38	Waste collection,treatment,disposal activities		...	...	...	...		...	...	...	...		...	...	...	...
39	Remediation activities		...	...	...	...		...	...	...	...		...	...	...	...
E	Water supply;sewerage,waste management		14	14	11	10		1920	1916	1904	1966		635	732	773	807

Mauritius

ISIC Revision 4			Output at basic prices					Value added at basic prices			
			(millions of Mauritian Rupees)					(millions of Mauritian Rupees)			
ISIC	Industry	Note	2012	2013	2014	2015	Note	2012	2013	2014	2015
05	Mining of coal and lignite		...	...	...	...		...	...	...	...
06	Extraction of crude petroleum & natural gas		...	...	...	...		...	...	...	...
07	Mining of metal ores		...	...	...	...		...	...	...	...
08	Other mining and quarrying		...	...	...	...		...	...	...	...
09	Mining support service activities		...	...	...	...		...	...	...	...
B	Mining and quarrying		2861	2576	2502	2586		1000	990	1000	893
35	Electricity, gas, steam & air conditioning		19285	19546	22512	28933		4306	4722	5511	7083
D	Electricity, gas, steam & air conditioning		19285	19546	22512	28933		4306	4722	5511	7083
36	Water collection, treatment and supply		...	...	...	...		...	...	...	...
37	Sewerage		...	...	...	...		...	...	...	...
38	Waste collection,treatment,disposal activities		...	...	...	...		...	...	...	...
39	Remediation activities		...	...	...	...		...	...	...	...
E	Water supply;sewerage,waste management		2130	2221	2368	2548		1220	1294	1340	1442

Mexico

ISIC Revision 4		Establishments					Number of employees					Wages and salaries paid to employees				
			(number)					(number)					(millions of Mexican Pesos)			
ISIC	Industry	Note	2012	2013	2014	2015	Note	2012	2013	2014	2015	Note	2012	2013	2014	2015
051	Mining of hard coal		...	...	...	...		...	6621	...	...		...	305	...	...
052	Mining of lignite		...	...	...	...		...	-	...	...		...	-	...	...
061	Extraction of crude petroleum		...	...	...	...		...	53581	...	...		...	20538	...	...
062	Extraction of natural gas		...	...	...	...		...	-	...	...		...	-	...	...
071	Mining of iron ores		...	...	...	...		...	4738	...	...		...	337	...	...
072	Mining of non-ferrous metal ores		...	...	...	...		...	25166	...	...		...	959	...	...
081	Quarrying of stone, sand and clay		...	...	...	...		...	14815	...	...		...	190	...	...
089	Mining and quarrying n.e.c.		...	...	...	...		...	7473	...	...		...	301	...	...
091	Support activities for petroleum		...	...	...	...		...	1363	...	...		...	77	...	...
099	Support activities for other mining, quarrying		...	...	...	...		...	2021	...	...		...	45	...	...
B	Mining and quarrying		...	...	...	...		...	115778	...	...		...	22752	...	...
351	Electric power generation, transmission		...	...	...	...		...	95699	...	...		...	...	...	...
352	Manufacture of gas		...	...	...	...		...	...	...	...		...	...	...	...
353	Steam and air conditioning supply		...	...	...	...		...	7755	...	...		...	...	...	...
D	Electricity, gas, steam & air conditioning		...	...	...	...		...	103454	...	...		...	...	...	...
360	Water collection, treatment and supply		...	...	...	...		...	114102	...	...		...	5184	...	...
370	Sewerage		...	...	...	...		...	-	...	...		...	-	...	...
381	Waste collection		...	...	...	...		...	11775	...	...		...	193	...	...
382	Waste treatment and disposal		...	...	...	...		...	5097	...	...		...	56	...	...
383	Materials recovery		...	...	...	...		...	-	...	...		...	-	...	...
390	Remediation activities		...	...	...	...		...	-	...	...		...	-	...	...
E	Water supply;sewerage,waste management		...	...	...	...		...	...	...	...		...	...	...	...

Mexico

ISIC Revision 4		Output (valuation not defined)					Value added (valuation not defined)				
		Note	(millions of Mexican Pesos)				Note	(millions of Mexican Pesos)			
ISIC	Industry		2012	2013	2014	2015		2012	2013	2014	2015
051	Mining of hard coal		...	8350	...	...		...	4379	...	...
052	Mining of lignite		...	-	...	...		...	-	...	...
061	Extraction of crude petroleum		...	1036357	...	...		...	895937	...	...
062	Extraction of natural gas		...	-	...	...		...	-	...	...
071	Mining of iron ores		...	11862	...	63243		...	4728	...	...
072	Mining of non-ferrous metal ores		...	142091	...	106339		...	88031	...	...
081	Quarrying of stone, sand and clay		...	8337	...	14234a/		...	3303	...	...
089	Mining and quarrying n.e.c.		...	8350	...	...a/		...	3638	...	...
091	Support activities for petroleum		...	9490	...	...		...	4110	...	...
099	Support activities for other mining, quarrying		...	2773	...	...		...	1396	...	...
B	Mining and quarrying		...	1227610	...	...		...	1005522	...	...
351	Electric power generation, transmission		...	371865	...	...		...	204586	...	...
352	Manufacture of gas		...	3384	...	...		...	998	...	...
353	Steam and air conditioning supply		...	3388	...	...		...	1544	...	...
D	Electricity, gas, steam & air conditioning		...	378638	...	...		...	207128	...	...
360	Water collection, treatment and supply		...	77878	...	...		...	52942	...	...
370	Sewerage		...	-	...	...		...	-	...	...
381	Waste collection		...	6594	...	...		...	3138	...	...
382	Waste treatment and disposal		...	1674	...	...		...	775	...	...
383	Materials recovery		...	-	...	...		...	-	...	...
390	Remediation activities		...	-	...	...		...	-	...	...
E	Water supply;sewerage,waste management		...	...	...	...		...	...	...	...

a/ 081 includes 089.

Mongolia

ISIC Revision 4		Number of establishments					Number of employees					Wages and salaries paid to employees				
			(number)					(number)					(millions of Mongolian Togrog)			
ISIC	Industry	Note	2012	2013	2014	2015	Note	2012	2013	2014	2015	Note	2012	2013	2014	2015
05	Mining of coal and lignite		...	...	81	73		7093	8847	8210	15571		120134	109017	107891	107008
051	Mining of hard coal		...	...	62	58		...	...	7233	13382		...	...	...	...
052	Mining of lignite		...	...	19	15		...	...	977	2189		...	...	...	...
06	Extraction of crude petroleum & natural gas		9	13	2	3		414	726	580	705		6952	8445	11294	9594
061	Extraction of crude petroleum		...	...	2	3		...	...	580	705		...	...	...	...
062	Extraction of natural gas		...	...	...	...		...	...	...	...		...	...	...	...
07	Mining of metal ores		187	174	182	194		12981	11881	14766	22992		239988	517099	562456	558533
071	Mining of iron ores		...	...	27	27		...	...	1325	4120		...	...	...	...
072	Mining of non-ferrous metal ores		...	...	155	167		...	...	13441	18872		...	...	...	...
08	Other mining and quarrying		135	115	136	145		2917	3334	2911	4811		25746	25672	25487	24990
081	Quarrying of stone, sand and clay		...	...	53	62		...	...	448	562		...	...	...	...
089	Mining and quarrying n.e.c.		...	...	83	83		...	...	2463	4249		...	...	...	...
09	Mining support service activities		329	318	470	479		6667	10143	6918	19254		396258	118042	58652	...
091	Support activities for petroleum		...	...	19	17		...	...	538	1329		...	...	...	...
099	Support activities for other mining, quarrying		...	...	451	462		...	...	6380	17925		...	...	...	...
B	Mining and quarrying		660	620	871	894		30072	34931	33385	63333		789078	778275	765780	740024
35	Electricity, gas, steam & air conditioning		187	178	212	192		13157	13495	14771	16457		138349	168927	184235	217559
351	Electric power generation, transmission		...	...	84	57		...	...	9937	11141		...	...	...	...
352	Manufacture of gas		...	...	...	...		...	...	...	...		...	...	...	...
353	Steam and air conditioning supply		...	...	128	135		...	...	4834	5316		...	...	...	...
D	Electricity, gas, steam & air conditioning		187	178	212	192		13157	13495	14771	16457		138349	168927	184235	217559
36	Water collection, treatment and supply		53	83	83	59		5333	4925	3399	3881		33972	34307	39516	57558a/
37	Sewerage		75	34	24	13		2126	863	819	519		1182	1468	639	...a/
38	Waste collection,treatment,disposal activities		25	49	45	72		100	1243	637	1925		10213	15799	...	...
381	Waste collection		...	...	21	52		...	...	435	1789		...	...	...	...
382	Waste treatment and disposal		...	...	13	10		...	...	135	105		...	...	...	...
383	Materials recovery		...	...	11	10		...	...	67	31		...	...	...	...
39	Remediation activities		16	7	21	18		339	12	509	433		90	63	1706	...
E	Water supply;sewerage,waste management		169	173	173	162		7898	7043	5364	6758		45457	51637	41861	87050

a/ 36 includes 37.

Mongolia

ISIC Revision 4		Output at basic prices					Value added at basic prices				
			(millions of Mongolian Togrog)					(millions of Mongolian Togrog)			
ISIC	Industry	Note	2012	2013	2014	2015	Note	2012	2013	2014	2015
05	Mining of coal and lignite		1722628	1461125	1081738	952492		665373	593788	383903	328191
051	Mining of hard coal		...	...	...	...		...	...	...	...
052	Mining of lignite		...	...	...	...		...	...	...	...
06	Extraction of crude petroleum & natural gas		430656	770940	1125185	834989		163337	305561	442283	290813
061	Extraction of crude petroleum		...	...	...	...		...	...	...	...
062	Extraction of natural gas		...	...	...	...		...	...	...	...
07	Mining of metal ores		2255636	3273706	6467457	5863579		1274962	1740617	2636387	3166538
071	Mining of iron ores		...	...	...	...		...	...	...	...
072	Mining of non-ferrous metal ores		...	...	...	...		...	...	...	...
08	Other mining and quarrying		154078	113897	137919	141855		45209	36107	39990	39902
081	Quarrying of stone, sand and clay		...	...	...	...		...	...	...	...
089	Mining and quarrying n.e.c.		...	...	...	...		...	...	...	...
09	Mining support service activities		1224027	480088	458987	361648		593869	173342	158467	130519
091	Support activities for petroleum		...	...	...	...		...	...	...	...
099	Support activities for other mining, quarrying		...	...	...	...		...	...	...	...
B	Mining and quarrying		5787025	6099756	9271286	8154563		2742750	2849415	3661030	3955963
35	Electricity, gas, steam & air conditioning		824143	904495	842769	1047366		233859	270514	313587	379867
351	Electric power generation, transmission		...	...	...	...		...	...	...	...
352	Manufacture of gas		...	...	...	...		...	...	...	...
353	Steam and air conditioning supply		...	...	...	...		...	...	...	...
D	Electricity, gas, steam & air conditioning		824143	904495	842769	1047366		233859	270514	313587	379867
36	Water collection, treatment and supply		135844	146020	141295	160763a/		52066	45796	62188	71194a/
37	Sewerage		10171	5643	7453	...a/		1702	1870	2587	...a/
38	Waste collection,treatment,disposal activities		50470	73944	...	...		18302	29607	...	...
381	Waste collection		...	...	...	...		...	...	...	...
382	Waste treatment and disposal		...	...	...	...		...	...	...	...
383	Materials recovery		...	...	...	...		...	...	...	...
39	Remediation activities		1626	1068	6446	120334		460	262	1915	32762
E	Water supply;sewerage,waste management		198111	226675	155194	281097		72530	77535	66690	103956

a/ 36 includes 37.

Montenegro

ISIC Revision 4		Establishments					Number of employees					Wages and salaries paid to employees				
			(number)					(number)					(thousands of Euros)			
ISIC	Industry	Note	2012	2013	2014	2015	Note	2012	2013	2014	2015	Note	2012	2013	2014	2015
051	Mining of hard coal		...	...	...	...		...	982a/	918a/	1010a/		...	...	11875a/	12811a/
052	Mining of lignite		...	...	...	...		...	...a/	...a/	...a/		...	...	...a/	...a/
061	Extraction of crude petroleum		...	...	...	...		...	...	...	...		...	...	...	...
062	Extraction of natural gas		...	...	...	...		...	...	...	...		...	...	...	...
071	Mining of iron ores		...	...	...	...		...	412b/	445b/	351b/		...	...	2948b/	2502b/
072	Mining of non-ferrous metal ores		...	...	...	...		...	...b/	...b/	...b/		...	...	...b/	...b/
081	Quarrying of stone, sand and clay		...	...	...	...		...	308c/	359c/	371c/		...	...	1827c/	1643c/
089	Mining and quarrying n.e.c.		...	...	...	...		...	...c/	...c/	...c/		...	...	...c/	...c/
091	Support activities for petroleum		...	...	...	...		...	172d/	114d/	73d/		...	...	10d/	...
099	Support activities for other mining, quarrying		...	...	...	...		...	...d/	...d/	...d/		...	...	...d/	...
B	Mining and quarrying		...	...	...	...		...	1874	1836	1805		...	...	20005	18606
351	Electric power generation, transmission		...	...	...	...		...	3031e/	2884e/	2826e/		...	...	42706e/	44459e/
352	Manufacture of gas		...	...	...	...		...	...e/	...e/	...e/		...	...	...e/	...e/
353	Steam and air conditioning supply		...	...	...	...		...	...e/	...e/	...e/		...	...	...e/	...e/
D	Electricity, gas, steam & air conditioning		...	...	...	...		...	3031	2884	2826		...	...	42706	44459
360	Water collection, treatment and supply		...	...	...	...		...	2053	2122	2183		...	...	18283	19044
370	Sewerage		...	...	...	...		...	8	17	8		...	...	...	...
381	Waste collection		...	...	...	...		...	217f/	1214f/	1213f/		...	...	9921f/	9796f/
382	Waste treatment and disposal		...	...	...	...		...	...f/	...f/	...f/		...	...	...f/	...f/
383	Materials recovery		...	...	...	...		...	...f/	...f/	...f/		...	...	...f/	...f/
390	Remediation activities		...	...	...	...		...	2495	1717	1647		...	...	13393	12787
E	Water supply;sewerage,waste management		...	...	...	...		...	4773	5070	5051		...	...	41797	41944

a/ 051 includes 052.
b/ 071 includes 072.
c/ 081 includes 089.
d/ 091 includes 099.
e/ 351 includes 352 and 353.
f/ 381 includes 382 and 383.

Montenegro

ISIC Revision 4		Output at basic prices					Value added at basic prices				
			(thousands of Euros)					(thousands of Euros)			
ISIC	Industry	Note	2012	2013	2014	2015	Note	2012	2013	2014	2015
051	Mining of hard coal		...	44368a/	44853a/	50577a/		...	28009a/	30909a/	33490a/
052	Mining of lignite		...	...a/	...a/	...a/		...	...a/	...a/	...a/
061	Extraction of crude petroleum		...	...	...	...		...	...	...	...
062	Extraction of natural gas		...	...	...	...		...	...	...	...
071	Mining of iron ores		...	11089b/	11933b/	11741b/		...	3209b/	5119b/	4303b/
072	Mining of non-ferrous metal ores		...	...b/	...b/	...b/		...	...b/	...b/	...b/
081	Quarrying of stone, sand and clay		...	9453c/	9269c/	7942c/		...	4702c/	3723c/	3728c/
089	Mining and quarrying n.e.c.		...	...c/	...c/	...c/		...	...c/	...c/	...c/
091	Support activities for petroleum		...	103d/	...	...		...	66d/	11d/	2d/
099	Support activities for other mining, quarrying		...	...d/	...	...		...	...d/	...d/	...d/
B	Mining and quarrying		...	65012	66064	70260		...	35986	39758	41521
351	Electric power generation, transmission		...	256096e/	239660e/	231800e/		...	156836e/	138749e/	126112e/
352	Manufacture of gas		...	...e/	...e/	...e/		...	...e/	...e/	...e/
353	Steam and air conditioning supply		...	...e/	...e/	...e/		...	...e/	...e/	...e/
D	Electricity, gas, steam & air conditioning		...	256096	239660	231800		...	156836	138749	126112
360	Water collection, treatment and supply		...	47014	43508	47358		...	34901	30664	37468
370	Sewerage		...	3	...	58		...	3	6	3
381	Waste collection		...	25342f/	21659f/	14270f/		...	10383f/	9163f/	8970f/
382	Waste treatment and disposal		...	...f/	...f/	...f/		...	...f/	...f/	...f/
383	Materials recovery		...	...f/	...f/	...f/		...	...f/	...f/	...f/
390	Remediation activities		...	30135	32196	31249		...	21775	24128	23612
E	Water supply;sewerage,waste management		...	102495	97364	93012		...	67061	63955	70109

a/ 051 includes 052.
b/ 071 includes 072.
c/ 081 includes 089.
d/ 091 includes 099.
e/ 351 includes 352 and 353.
f/ 381 includes 382 and 383.

Netherlands

ISIC Revision 4		Number of enterprises (number)					Number of employees (number)					Wages and salaries paid to employees (millions of Euros)			
ISIC	Industry	Note	2012	2013	2014	2015	Note	2012	2013	2014	2015	Note	2012	2013	2014	2015
05	Mining of coal and lignite		-	-	-	-		-	-	-	-		-	-	-	-
051	Mining of hard coal		-	-	-	-		-	-	-	-		-	-	-	-
052	Mining of lignite		-	-	-	-		-	-	-	-		-	-	-	-
06	Extraction of crude petroleum & natural gas		33	45	43	41		3376	3615	3818	3903		310	327	356	391
061	Extraction of crude petroleum		26	34	33	31		625	577	727	660	a/	310	327	356	391
062	Extraction of natural gas		7	11	10	10		2751	3038	3091	3243	a/	...	...	...	...
07	Mining of metal ores		-	-	-	-		-	-	-	-		-	-	-	-
071	Mining of iron ores		-	-	-	-		-	-	-	-		-	-	-	-
072	Mining of non-ferrous metal ores		-	-	-	-		-	-	-	-		-	-	-	-
08	Other mining and quarrying		144	162	157	159		2092	2122	1849	1851		83	81	84	92
081	Quarrying of stone, sand and clay		114	128	123	126		781	786	756	724		38	36	38	42
089	Mining and quarrying n.e.c.		30	34	34	33		1311	1336	1093	1128		45	45	47	50
09	Mining support service activities		144	210	237	264		3125	4528	5530	5399		268	365	393	467
091	Support activities for petroleum		139	201	228	251	b/	3125	4528	5530	5399		...	365b/	393b/	...
099	Support activities for other mining, quarrying		5	9	9	13	b/	...	...	...	...		-	...b/	...b/	1
B	Mining and quarrying		321	417	437	464		8593	10265	11197	11153		661	773	834	950
35	Electricity, gas, steam & air conditioning		765	950	1191	1130		25799	26241	26340	26803		1419	1530	1561	1610
351	Electric power generation, transmission		757	940	1176	1115	c/	25799	26241	26340	26803		1415	1529	1561	1609
352	Manufacture of gas		6	6	11	12	c/	...	...	...	...		...	...	...	...
353	Steam and air conditioning supply		2	4	4	3	c/	...	...	...	...		...	...	...	...
D	Electricity, gas, steam & air conditioning		765	950	1191	1130		25799	26241	26340	26803		1419	1530	1561	1610
36	Water collection, treatment and supply		24	29	31	29		5291	5329	5343	5327		272	282	279	290
37	Sewerage		139	138	140	138		4483	3867	3853	3790		165	171	170	165
38	Waste collection,treatment,disposal activities		806	892	886	936		26321	21535	21774	21434		849	884	877	885
381	Waste collection		288	269	258	251		15840	10967	10421	10122		399	402	372	380
382	Waste treatment and disposal		217	247	238	232		6211	6258	6974	7066		271	303	323	314
383	Materials recovery		301	376	390	453		4270	4310	4379	4246		180	179	182	191
39	Remediation activities		250	386	460	578		1604	1613	1577	1610		74	67	66	68
E	Water supply;sewerage,waste management		1219	1445	1517	1681		37699	32343	32546	32160		1361	1405	1392	1407

a/ 061 includes 062.
b/ 091 includes 099.
c/ 351 includes 352 and 353.

Netherlands

ISIC Revision 4		Output (valuation not defined)					Value added at factor values				
			(millions of Euros)					(millions of Euros)			
ISIC	Industry	Note	2012	2013	2014	2015	Note	2012	2013	2014	2015
05	Mining of coal and lignite		-	-	-	-		-	-	-	-
051	Mining of hard coal		-	-	-	-		-	-	-	-
052	Mining of lignite		-	-	-	-		-	-	-	-
06	Extraction of crude petroleum & natural gas		21194	21855	16891	12079		9993	9547	7829	6120
061	Extraction of crude petroleum	a/	21194	21855	16891	12079	a/	9993	9547	7829	6120
062	Extraction of natural gas	a/	...	...	...	...	a/	...	...	...	...
07	Mining of metal ores		-	-	-	-		-	-	-	-
071	Mining of iron ores		-	-	-	-		-	-	-	-
072	Mining of non-ferrous metal ores		-	-	-	-		-	-	-	-
08	Other mining and quarrying		934	858	937	1023		233	207	237	271
081	Quarrying of stone, sand and clay		431	370	418	438		114	110	126	145
089	Mining and quarrying n.e.c.		503	488	519	586		119	97	111	126
09	Mining support service activities		2806	3449	3757	4768		1098	1295	1204	1747
091	Support activities for petroleum		...	3449b/	3757b/	...		...	1295b/	1204b/	...
099	Support activities for other mining, quarrying		-	...b/	...b/	2		-	...b/	...b/	1
B	Mining and quarrying		24933	26162	21585	17870		11325	11048	9270	8138
35	Electricity, gas, steam & air conditioning		27396	25572	20274	17653		7357	8170	6386	6937
351	Electric power generation, transmission		27394	25563	20259	17639		7366	8168	6383	6937
352	Manufacture of gas		...	...	...	...		...	...	...	...
353	Steam and air conditioning supply		...	...	...	...		...	...	...	...
D	Electricity, gas, steam & air conditioning		27396	25572	20274	17653		7357	8170	6386	6937
36	Water collection, treatment and supply		1477	1500	1475	1530		951	977	987	995
37	Sewerage		649	663	689	654		258	267	273	251
38	Waste collection,treatment,disposal activities		6742	6700	6809	6920		2131	2093	2209	2220
381	Waste collection		2158	2093	1995	2061		754	729	699	725
382	Waste treatment and disposal		1848	2069	2322	2257		901	968	1093	1064
383	Materials recovery		2736	2539	2492	2602		477	395	417	431
39	Remediation activities		409	429	378	399		143	142	128	136
E	Water supply;sewerage,waste management		9277	9292	9350	9503		3482	3478	3597	3602

a/ 061 includes 062.
b/ 091 includes 099.

New Zealand

ISIC Revision 4		Number of enterprises					Number of persons engaged					Wages and salaries paid to employees				
			(number)					(number)					(millions of New Zealand Dollars)			
ISIC	Industry	Note	2012	2013	2014	2015	Note	2012	2013	2014	2015	Note	2012	2013	2014	2015
051	Mining of hard coal		18	21	15	15		1600	1250	1150	820		...	...	...	...
052	Mining of lignite		-	-	-	-		-	-	-	-		...	...	...	...
061	Extraction of crude petroleum		6	12	12	12		470	870	920	860		...	...	173	...
062	Extraction of natural gas		-	-	-	-		-	-	-	-		...	...	-	...
071	Mining of iron ores		3	3	3	3		160	170	220	230	a/	481	120	112	128
072	Mining of non-ferrous metal ores		39	36	36	36		1350	1200	1150	1000	a/	...	...	...	...
081	Quarrying of stone, sand and clay		117	111	120	117		1000	1050	1150	1200		109b/	100	107	110
089	Mining and quarrying n.e.c.		6	9	9	9		120	180	140	180		...b/	...	...	...
091	Support activities for petroleum		63	63	63	60		1050	1100	870	730		134c/	124	135	129
099	Support activities for other mining, quarrying		9	9	9	9		75	95	40	40		...c/	12	6	4
B	Mining and quarrying		261	264	267	261		5825	5915	5640	5060		724	726	665	685
351	Electric power generation, transmission		57	57	63	69		9800	9900	10000	10100	d/	787	784	818	855
352	Manufacture of gas		3	6	3	3		55	250	280	290	d/	...	...	...	...
353	Steam and air conditioning supply		-	-	-	-		-	-	-	-	d/	...	...	...	...
D	Electricity, gas, steam & air conditioning		60	63	66	72		9855	10150	10280	10390		787	784	818	855
360	Water collection, treatment and supply		27	30	30	30		850	890	1050	1100		...	...	...	...
370	Sewerage		6	9	15	12		12	30	45	65		...	...	...	...
381	Waste collection		195	189	198	204		2350	2150	2150	2250		129e/	88	105	99
382	Waste treatment and disposal		78	81	78	78		580	720	860	920		...e/	71	45	48
383	Materials recovery		-	-	-	-		-	-	-	-		...e/	-	-	-
390	Remediation activities		84	90	99	93		1850	1800	2150	2200		97	97	105	114
E	Water supply;sewerage,waste management		390	399	420	417		5642	5590	6255	6535		275	304	317	336

a/ 071 includes 072.
b/ 081 includes 089.
c/ 091 includes 099.
d/ 351 includes 352 and 353.
e/ 381 includes 382 and 383.

New Zealand

ISIC Revision 4		Output (valuation not defined)					Value added (valuation not defined)				
			(millions of New Zealand Dollars)					(millions of New Zealand Dollars)			
ISIC	Industry	Note	2012	2013	2014	2015	Note	2012	2013	2014	2015
051	Mining of hard coal		...	...	...	...		...	...	...	...
052	Mining of lignite		...	...	...	...		...	...	...	...
061	Extraction of crude petroleum		...	...	4730	...		...	...	2070	...
062	Extraction of natural gas		...	...	-	...		...	...	-	...
071	Mining of iron ores		...	942a/	974a/	887a/		...	...	421a/	354a/
072	Mining of non-ferrous metal ores		...	...a/	...a/	...a/		...	...	...a/	...a/
081	Quarrying of stone, sand and clay		643b/	588	674	703		285b/	258	287	310
089	Mining and quarrying n.e.c.		...b/	...	...	...		...b/	...	...	...
091	Support activities for petroleum		732c/	636	753	970		225c/	259	333	449
099	Support activities for other mining, quarrying		...c/	78	45	30		...c/	-8	7	9
B	Mining and quarrying		7778	7697	7852	8629		3410	3376	3300	3374
351	Electric power generation, transmission	d/	16214	16134	15491	16438	d/	5423	5395	5599	5686
352	Manufacture of gas	d/	...	...	...	...	d/	...	...	...	...
353	Steam and air conditioning supply	d/	...	...	...	...	d/	...	...	...	...
D	Electricity, gas, steam & air conditioning		16214	16134	15491	16438		5423	5395	5599	5686
360	Water collection, treatment and supply		...	...	...	...		...	...	...	...
370	Sewerage		...	...	...	...		...	...	...	...
381	Waste collection		678e/	334	480	482		311e/	164	233	241
382	Waste treatment and disposal		...e/	406	280	308		...e/	201	127	143
383	Materials recovery		...e/	-	...	-		...e/	-	-	-
390	Remediation activities		525	525	525	571		...	224	219	272
E	Water supply;sewerage,waste management		1748	1810	1848	1978		906	959	970	1100

a/ 071 includes 072.
b/ 081 includes 089.
c/ 091 includes 099.
d/ 351 includes 352 and 353.
e/ 381 includes 382 and 383.

Niger

ISIC Revision 3		Number of enterprises					Number of persons engaged					Wages and salaries paid to employees				
			(number)					(number)					(millions of West African CFA Francs)			
ISIC	Industry	Note	2012	2013	2014	2015	Note	2012	2013	2014	2015	Note	2012	2013	2014	2015
10	Mining of coal and lignite; extraction of peat		...	...	...	...		...	...	...	...		...	...	...	...
11	Extraction of crude oil and natural gas, etc.		...	...	...	...		...	...	...	...		...	...	...	...
12	Mining of uranium and thorium ores		...	...	...	...		...	...	...	...		...	...	...	...
13	Mining of metal ores		...	...	...	...		...	...	...	...		...	...	...	...
14	Other mining and quarrying		...	...	...	...		...	...	...	...		...	...	...	...
C	Total mining and quarrying		...	...	6	6		...	...	19262	19828		...	...	56057	55781
40	Electricity, gas, steam and hot water supply		...	...	...	...		...	...	...	...		...	...	...	...
41	Water collection, purification, distribution		...	...	...	...		...	...	...	...		...	...	...	...
E	Total electricity, gas and water supply		...	...	5	5		...	...	8841	9433		...	...	17520	18504

Niger

ISIC Revision 3		Output (valuation not defined)					Value added (valuation not defined)				
			(millions of West African CFA Francs)					(millions of West African CFA Francs)			
ISIC	Industry	Note	2012	2013	2014	2015	Note	2012	2013	2014	2015
10	Mining of coal and lignite; extraction of peat		...	...	...	...		...	...	...	...
11	Extraction of crude oil and natural gas, etc.		...	...	...	...		...	...	...	...
12	Mining of uranium and thorium ores		...	...	...	...		...	...	...	...
13	Mining of metal ores		...	...	...	...		...	...	...	...
14	Other mining and quarrying		...	...	...	...		...	...	...	...
C	Total mining and quarrying		...	...	598595	559704		...	...	364517	329959
40	Electricity, gas, steam and hot water supply		...	...	...	...		...	...	...	...
41	Water collection, purification, distribution		...	...	...	...		...	...	...	...
E	Total electricity, gas and water supply		...	...	117534	121582		...	...	49830	51546

Norway

ISIC Revision 4		Number of enterprises					Number of employees					Wages and salaries paid to employees				
			(number)					(number)					(millions of Norwegian Kroner)			
ISIC	Industry	Note	2012	2013	2014	2015	Note	2012	2013	2014	2015	Note	2012	2013	2014	2015
051	Mining of hard coal		1	1	1	1		402	378	347	219		280	279	271	214
052	Mining of lignite		-	-	-	-		-	-	-	-		-	-	-	-
061	Extraction of crude petroleum		68	71	67	56		26452	27297	27138	25890		...	28477	29335	...
062	Extraction of natural gas		5	4	4	4		1350	1423	1427	1343		...	1878	1483	...
071	Mining of iron ores		6	6	5	4		612	644	659	648		...	...	...	...
072	Mining of non-ferrous metal ores		3	4	3	4		298	307	307	301		...	...	...	...
081	Quarrying of stone, sand and clay		610	623	629	646		2996	2978	2911	3010		1342	1365	1381	1379
089	Mining and quarrying n.e.c.		38	30	32	31		539	406	378	391		256	212	201	206
091	Support activities for petroleum		332	332	353	355		33054	36355	37916	35015		22160	28804	31179	29416
099	Support activities for other mining, quarrying		17	17	15	20		30	18	16	21		13	10	11	12
B	Mining and quarrying		1080	1088	1109	1121		65733	69806	71099	66838		52687	61597	64418	...
351	Electric power generation, transmission		951	962	952	294		14828	15143	15252	12694		8576	8170	8352	...
352	Manufacture of gas		21	26	22	16		55	65	66	61		43	44	63	...
353	Steam and air conditioning supply		124	136	143	153		467	535	522	534		281	313	312	...
D	Electricity, gas, steam & air conditioning		1096	1124	1117	463		15350	15743	15840	13289		8900	8526	8727	...
360	Water collection, treatment and supply		...	...	...	...		320	343	377	397		84	105	87	90
370	Sewerage		157	165	170	173		678	684	729	760		315	319	343	360
381	Waste collection		230	237	237	237		4220	4263	4540	4676		1733	1796	1924	1999
382	Waste treatment and disposal		69	72	72	73		771	846	528	683		404	455	246	296
383	Materials recovery		131	121	122	126		1787	1756	2063	2130		851	842	1069	1078
390	Remediation activities		35	36	32	30		80	53	61	63		45	22	31	33
E	Water supply;sewerage,waste management		1084	1132	1156	1173		7856	7945	8298	8709		3430	3540	3700	3857

Norway

ISIC Revision 4		Output (valuation not defined)					Value added at factor values				
			(millions of Norwegian Kroner)					(millions of Norwegian Kroner)			
ISIC	Industry	Note	2012	2013	2014	2015	Note	2012	2013	2014	2015
051	Mining of hard coal		851	1171	964	525		185	421	-145	14
052	Mining of lignite		-	-	-	-		-	-	-	-
061	Extraction of crude petroleum		...	1103261	1052851	497660		...	513581	455890	407148
062	Extraction of natural gas		...	48877	44946	57886		...	27227	31420	51236
071	Mining of iron ores		...	...	...	...		...	...	...	...
072	Mining of non-ferrous metal ores		...	...	...	...		...	...	...	...
081	Quarrying of stone, sand and clay		8208	8390	8792	8977		2744	2893	3184	3363
089	Mining and quarrying n.e.c.		1589	1476	1541	1634		658	591	608	662
091	Support activities for petroleum		111360	132190	139336	137495		39536	48707	51831	50719
099	Support activities for other mining, quarrying		47	35	41	46		29	20	23	17
B	Mining and quarrying		1379003	1298354	1251157	706221		651814	594618	543666	514137
351	Electric power generation, transmission		106363	118364	109014	66658		45240	51484	45237	46652
352	Manufacture of gas		632	728	900	558		14	-504	-222	47
353	Steam and air conditioning supply		3104	3688	3242	3445		1114	1395	1262	1516
D	Electricity, gas, steam & air conditioning		110100	122781	113155	70660		46368	52375	46276	48214
360	Water collection, treatment and supply		391	399	449	473		168	161	221	203
370	Sewerage		1118	1165	1270	1326		573	585	626	651
381	Waste collection		9365	9767	10333	10675		3240	3167	3348	3439
382	Waste treatment and disposal		2516	2709	1686	2012		980	1077	674	798
383	Materials recovery		8322	7879	8522	9087		1949	1740	2031	2292
390	Remediation activities		317	117	109	166		163	59	63	73
E	Water supply;sewerage,waste management		22029	22035	22370	23737		7072	6789	6963	7456

Peru

ISIC Revision 4		Output at basic prices					Value added at basic prices				
		Note	(millions of Peruvian New Soles)				Note	(millions of Peruvian New Soles)			
ISIC	Industry		2012	2013	2014	2015		2012	2013	2014	2015
05	Mining of coal and lignite		...	...	...	...		...	...	...	...
06	Extraction of crude petroleum & natural gas		23080	22845	21417	12927		13131	12388	10655	6387
07	Mining of metal ores		70145	66808	61413	65479		48651	44232	39091	41783
08	Other mining and quarrying		...	...	...	...		...	...	...	...
09	Mining support service activities		...	...	...	...		...	...	...	...
B	Mining and quarrying		93225	89653	82830	78406		61782	56620	49746	48170
35	Electricity, gas, steam & air conditioning		12296	13516	14924	17445		6598	7269	8492	9923
D	Electricity, gas, steam & air conditioning		12296a/	13516	14924	17445		6598	7269	8492	9923
36	Water collection, treatment and supply		3188	3383	3601	4096		2003	2086	2226	2528
37	Sewerage		...	...	...	...		...	...	...	...
38	Waste collection,treatment,disposal activities		...	...	...	...		...	...	...	...
39	Remediation activities		...	...	...	...		...	...	...	...
E	Water supply;sewerage,waste management		3188a/	3383	3601	4096		2013	2086	2226	2528

a/ Sum of available data.

Philippines

ISIC Revision 4		Number of establishments					Number of employees					Wages and salaries paid to employees				
		Note	(number)				Note	(number)				Note	(millions of Philippine Pesos)			
ISIC	Industry		2012	2013	2014	2015		2012	2013	2014	2015		2012	2013	2014	2015
05	Mining of coal and lignite		11	10	11	...		3766	3167	3314	...		839	941	735	...
051	Mining of hard coal		11	10	11	...		3766	3167	3314	...		839	941	735	...
052	Mining of lignite		...	...	...	...		...	...	...	...		...	...	...	...
06	Extraction of crude petroleum & natural gas		7	8	6	...		508	591	514	...		867	1033	946	...
061	Extraction of crude petroleum		3	4	3	...		49	82	70	...		62	108	76	...
062	Extraction of natural gas		4	4	3	...		459	509	444	...		805	925	870	...
07	Mining of metal ores		102	85	85	...		28389	26512	28052	...		6326	6631	8769	...
071	Mining of iron ores		16	10	9	...		440	241	154	...		49	30	26	...
072	Mining of non-ferrous metal ores		86	75	76	...		27949	26271	27898	...		6277	6601	8743	...
08	Other mining and quarrying		167	124	130	...		4015	3200	3752	...		391	355	449	...
081	Quarrying of stone, sand and clay		137	106	111	...		3417	2637	3244	...		357	329	431	...
089	Mining and quarrying n.e.c.		30	18	19	...		598	563	508	...		34	26	19	...
09	Mining support service activities		26	20	24	...		1927	2102	1472	...		1025	815	825	...
091	Support activities for petroleum		6	5	5	...		575	438	329	...		466	385	351	...
099	Support activities for other mining, quarrying		20	15	19	...		1352	1664	1143	...		558	429	474	...
B	Mining and quarrying		317	253	256	...		38658	35869	37104	...		9452	9818	11724	...
35	Electricity, gas, steam & air conditioning		265	253	259	...		47973	46696	46881	...		26589	29305	28090	...
351	Electric power generation, transmission		259	253a/	259a/	...		46707	46696a/	46881a/	...		24653	29305a/	28090a/	...
352	Manufacture of gas		-	-	-	...		-	-	-	...		-	-	-	...
353	Steam and air conditioning supply		6	...a/	...a/	...		1266	...a/	...a/	...		1936	...a/	...a/	...
D	Electricity, gas, steam & air conditioning		265	253	259	...		47973	46696	46881	...		26589	29305	28090	...
36	Water collection, treatment and supply		720	842	845	...		27815	30080	29622	...		7673	8784	8782	...
37	Sewerage		11	17	24	...		162	337	614	...		32	40	121	...
38	Waste collection,treatment,disposal activities		90	106	97	...		2974	2808	2839	...		429	406	394	...
381	Waste collection		24	33	31	...		1603	1324	1386	...		255	199	174	...
382	Waste treatment and disposal		13	18	13	...		419	514	344	...		77	85	86	...
383	Materials recovery		53	55	53	...		952	970	1109	...		97	122	134	...
39	Remediation activities		7	12	9	...		174	179	74	...		34	41	26	...
E	Water supply;sewerage,waste management		828	977	975	...		31125	33404	33150	...		8168	9271	9323	...

a/ 351 includes 353.

Philippines

ISIC Revision 4		Output at producers' prices					Value added at producers' prices				
			(millions of Philippine Pesos)					(millions of Philippine Pesos)			
ISIC	Industry	Note	2012	2013	2014	2015	Note	2012	2013	2014	2015
05	Mining of coal and lignite		19942	17169	19090	...		5468	6931	5635	...
051	Mining of hard coal		19942	17169	19090	...		5468	6931	5635	...
052	Mining of lignite		...	...	...	...		...	...	...	...
06	Extraction of crude petroleum & natural gas		66793	66781	64463	...		57052	57259	55088	...
061	Extraction of crude petroleum		...	729	969	...		-166	-94	525	...
062	Extraction of natural gas		66793	66053	63494	...		57218	57353	54563	...
07	Mining of metal ores		93242	95835	116220	...		48247	45576	69054	...
071	Mining of iron ores		458	203	312	...		-84	49	22	...
072	Mining of non-ferrous metal ores		92783	95632	115909	...		48331	45527	69032	...
08	Other mining and quarrying		5234	4463	5815	...		1398	1344	1624	...
081	Quarrying of stone, sand and clay		5072	4360	5717	...		1340	1298	1593	...
089	Mining and quarrying n.e.c.		161	104	98	...		58	45	31	...
09	Mining support service activities		6097	6028	8213	...		2054	2973	4185	...
091	Support activities for petroleum		1929	2037	1308	...		705	959	-114	...
099	Support activities for other mining, quarrying		4167	3991	6904	...		1348	2014	4299	...
B	Mining and quarrying		191324	190512	213801	...		114213	114151	135586	...
35	Electricity, gas, steam & air conditioning		732926	749022	732176	...		213545	200592	189180	...
351	Electric power generation, transmission		723000	749022a/	732176a/	...		207709	200592a/	189180a/	...
352	Manufacture of gas		-	-	-	...		-	-	-	...
353	Steam and air conditioning supply		9926	...a/	...a/	...		5836	...a/	...a/	...
D	Electricity, gas, steam & air conditioning		732926	749022	732176	...		213545	200592	189180	...
36	Water collection, treatment and supply		55167	61918	58838	...		34384	42319	40427	...
37	Sewerage		181	213	856	...		22	92	236	...
38	Waste collection,treatment,disposal activities		4962	5108	5319	...		1290	1286	2186	...
381	Waste collection		3440	2301	2111	...		942	536	536	...
382	Waste treatment and disposal		1041	1496	559	...		157	582	302	...
383	Materials recovery		481	1311	2649	...		191	167	1347	...
39	Remediation activities		112	150	121	...		41	59	59	...
E	Water supply;sewerage,waste management		60422	67390	65135	...		35737	43756	42907	...

a/ 351 includes 353.

Poland

ISIC Revision 4		Number of enterprises					Number of employees					Wages and salaries paid to employees				
			(number)					(number)					(millions of Polish Zlotys)			
ISIC	Industry	Note	2012	2013	2014	2015	Note	2012	2013	2014	2015	Note	2012	2013	2014	2015
051	Mining of hard coal		51	51	53	57		116783	112104	106157	94283		9602	9422	9253	7783
052	Mining of lignite		8	6	5	5		5092	4619	...	1734		325	269	...	83
061	Extraction of crude petroleum		24	26	28	28		...	...	677a/	...		...	...	147a/	...
062	Extraction of natural gas		29	26	27	34		82	130	...a/	...		32	79	...a/	...
071	Mining of iron ores		-	-	2	2		-	-	...	...		-	-	...	...
072	Mining of non-ferrous metal ores		11	12	15	13		...	...	...	...		...	...	...	...
081	Quarrying of stone, sand and clay		1433	1177	1330	1443		17214	15605	14906	15130		801	721	726	738
089	Mining and quarrying n.e.c.		94	90	100	108		4409	4143	4146	3639		238	241	245	217
091	Support activities for petroleum		102	92	107	104		4429	5047	5073	4568		350	436	426	376
099	Support activities for other mining, quarrying		192	177	185	214		5933	8893	9920	12623		255	373	412	621
B	Mining and quarrying		1944	1657	1852	2008		173068	169764	162181	150903		13785	13697	13479	12013
351	Electric power generation, transmission		2015	1933	1960	2537		87213	81674	77447	72604		6868	6621	6273	6219
352	Manufacture of gas		92	74	76	92		...	...	22359	19096		...	...	1436	1500
353	Steam and air conditioning supply		623	539	547	563		...	...	33640	33408		...	...	2012	2062
D	Electricity, gas, steam & air conditioning		2730	2546	2583	3192		145324	137505	133446	125108		10542	10106	9721	9780
360	Water collection, treatment and supply		627	658	667	630		31620	32023	31822	31523		1457	1507	1554	1584
370	Sewerage		2081	2103	2078	1911		32351	33497	34266	34441		1486	1568	1672	1740
381	Waste collection		2063	1939	2111	2355		35307	36220	38068	37707		1347	1419	1518	1545
382	Waste treatment and disposal		358	345	382	461		8568	9626	10061	11226		352	401	436	496
383	Materials recovery		1414	1406	1546	1745		9286	8690	8595	8198		349	309	335	337
390	Remediation activities		246	233	276	340		1275	1142	1333	1572		48	41	51	63
E	Water supply;sewerage,waste management		6789	6684	7060	7442		118407	121198	124145	124667		5039	5245	5567	5765

a/ 061 includes 062.

Poland

ISIC Revision 4		Output (valuation not defined)					Value added at factor values				
			(millions of Polish Zlotys)					(millions of Polish Zlotys)			
ISIC	Industry	Note	2012	2013	2014	2015	Note	2012	2013	2014	2015
051	Mining of hard coal		30081	26961	25959	25308		19842	17044	16125	16751
052	Mining of lignite		1017	1060	...	931		583	664	...	405
061	Extraction of crude petroleum		...	...	824a/	...		...	...	431a/	...
062	Extraction of natural gas		71	130	...a/	...		15	-6	...a/	...
071	Mining of iron ores		-	-	...	...		-	-	...	...
072	Mining of non-ferrous metal ores		...	...	...	...		...	...	...	...
081	Quarrying of stone, sand and clay		6989	5999	5623	5777		2307	1836	1981	2201
089	Mining and quarrying n.e.c.		1430	1306	1096	1151		708	590	544	505
091	Support activities for petroleum		1319	1794	1690	1392		565	785	815	695
099	Support activities for other mining, quarrying		847	1458	1481	1765		429	981	1031	1607
B	Mining and quarrying		63549	57206	54323	53016		37694	32796	30669	31378
351	Electric power generation, transmission		71322	68711	68759	67799		37023	36376	35371	36469
352	Manufacture of gas		...	...	35072	37797		...	...	8932	8459
353	Steam and air conditioning supply		...	...	13200	13944		...	...	6472	7421
D	Electricity, gas, steam & air conditioning		116008	115440	117031	119540		52250	51572	50774	52348
360	Water collection, treatment and supply		6250	6452	6733	6876		4121	4307	4532	4664
370	Sewerage		6976	8654	7892	8153		4653	6157	5324	5638
381	Waste collection		7160	7790	8667	9204		3107	3204	3564	3727
382	Waste treatment and disposal		2118	2311	2522	3121		1022	1140	1237	1451
383	Materials recovery		5459	4681	5133	4619		991	766	963	854
390	Remediation activities		440	372	570	573		116	83	116	183
E	Water supply;sewerage,waste management		28403	30261	31518	32546		14009	15658	15736	16517

a/ 061 includes 062.

Portugal

ISIC Revision 4		Number of enterprises					Number of employees					Wages and salaries paid to employees				
			(number)					(number)					(millions of Euros)			
ISIC	Industry	Note	2012	2013	2014	2015	Note	2012	2013	2014	2015	Note	2012	2013	2014	2015
05	Mining of coal and lignite		-	-	-	-		-	-	-	-		-	-	-	-
051	Mining of hard coal		-	-	-	-		-	-	-	-		-	-	-	-
052	Mining of lignite		-	-	-	-		-	-	-	-		-	-	-	-
06	Extraction of crude petroleum & natural gas		-	-	-	-		-	-	-	-		-	-	-	-
061	Extraction of crude petroleum		-	-	-	-		-	-	-	-		-	-	-	-
062	Extraction of natural gas		-	-	-	-		-	-	-	-		-	-	-	-
07	Mining of metal ores		19	19	19	23		...	...	...	...		...	...	...	...
071	Mining of iron ores		4	3	5	6		...	...	...	...		...	...	...	...
072	Mining of non-ferrous metal ores		15	16	14	17		...	...	...	...		...	...	...	...
08	Other mining and quarrying		1138	1119	1061	1021		7694	7027	6705	6613		98	88	85	85
081	Quarrying of stone, sand and clay		1057	1026	945	912		7435	6759	6456	6362		95	85	82	82
089	Mining and quarrying n.e.c.		81	93	116	109		259	268	249	251		3	3	3	3
09	Mining support service activities		19	19	22	22		...	...	...	...		...	...	...	...
091	Support activities for petroleum		-	2	3	4		-	...	...	...		-	...	...	...
099	Support activities for other mining, quarrying		19	17	19	18		...	366	...	...		...	7	...	...
B	Mining and quarrying		1176	1157	1102	1066		9753	9091	8824	8731		155	145	144	146
35	Electricity, gas, steam & air conditioning		888	925	941	1209		8555	8163	7962	8563		293	317	304	343
351	Electric power generation, transmission		830	867	884	1147		7765	7437	7239	7841		263	291	277	315
352	Manufacture of gas		41	41	40	39		716	652	645	639		28	25	26	26
353	Steam and air conditioning supply		17	17	17	23		74	74	78	83		2	2	2	2
D	Electricity, gas, steam & air conditioning		888	925	941	1209		8555	8163	7962	8563		293	317	304	343
36	Water collection, treatment and supply		141	142	151	151		13534	13194	13597	13497		199	213	211	209
37	Sewerage		71	71	75	64		1702	1675	1677	1608		27	29	28	27
38	Waste collection,treatment,disposal activities		963	985	1005	1028		14762	14567	13973	14241		194	198	187	193
381	Waste collection		357	359	378	421		5807	5995	5940	5681		70	73	72	70
382	Waste treatment and disposal		109	113	107	106		5407	5188	4817	5283		78	81	73	81
383	Materials recovery		497	513	520	501		3548	3384	3216	3277		46	44	42	42
39	Remediation activities		24	26	21	19		68	71	60	52		1	1	1	-
E	Water supply;sewerage,waste management		1199	1224	1252	1262		30066	29507	29307	29398		420	441	426	430

Portugal

ISIC Revision 4		Output (valuation not defined)					Value added at factor values				
		Note	(millions of Euros)				Note	(millions of Euros)			
ISIC	Industry		2012	2013	2014	2015		2012	2013	2014	2015
05	Mining of coal and lignite		-	-	-	-		-	-	-	-
051	Mining of hard coal		-	-	-	-		-	-	-	-
052	Mining of lignite		-	-	-	-		-	-	-	-
06	Extraction of crude petroleum & natural gas		-	-	-	-		-	-	-	-
061	Extraction of crude petroleum		-	-	-	-		-	-	-	-
062	Extraction of natural gas		-	-	-	-		-	-	-	-
07	Mining of metal ores		...	...	...	...		...	...	...	...
071	Mining of iron ores		...	...	...	...		...	...	...	...
072	Mining of non-ferrous metal ores		...	...	...	...		...	...	...	...
08	Other mining and quarrying		538	501	481	493		185	182	170	185
081	Quarrying of stone, sand and clay		520	482	461	472		181	175	163	178
089	Mining and quarrying n.e.c.		18	19	20	21		5	7	7	7
09	Mining support service activities		...	...	...	...		...	...	...	...
091	Support activities for petroleum		-	...	...	...		-	...	...	...
099	Support activities for other mining, quarrying		...	49	...	...		...	28	...	...
B	Mining and quarrying		1104	1028	1000	994		462	419	414	389
35	Electricity, gas, steam & air conditioning		13414	13005	12439	10821		4280	4382	4494	4070
351	Electric power generation, transmission		12514	12118	11499	9936		3835	3920	3964	3623
352	Manufacture of gas		882	868	922	866		438	454	522	440
353	Steam and air conditioning supply		19	19	19	19		7	8	8	7
D	Electricity, gas, steam & air conditioning		13414	13005	12439	10821		4280	4382	4494	4070
36	Water collection, treatment and supply		1324	1316	1321	1419		707	707	743	766
37	Sewerage		241	259	268	365		134	147	152	201
38	Waste collection,treatment,disposal activities		1264	1193	1195	1224		506	483	458	471
381	Waste collection		309	308	313	302		143	143	146	136
382	Waste treatment and disposal		492	482	496	546		243	238	213	246
383	Materials recovery		464	403	387	377		120	102	100	89
39	Remediation activities		2	2	1	1		1	1	-	-
E	Water supply;sewerage,waste management		2831	2770	2785	3009		1348	1337	1354	1439

Qatar

ISIC Revision 4		Number of establishments					Number of employees					Wages and salaries paid to employees				
			(number)					(number)					(millions of Qatari Riyals)			
ISIC	Industry	Note	2012	2013	2014	2015	Note	2012	2013	2014	2015	Note	2012	2013	2014	2015
051	Mining of hard coal		...	...	...	...		...	...	...	...		...	...	...	...
052	Mining of lignite		...	...	...	...		...	...	...	...		...	...	...	...
061	Extraction of crude petroleum		...	...	7a/	7a/		...	...	20212a/	18651a/		...	...	10319a/	10770a/
062	Extraction of natural gas		...	...	...a/	...a/		...	...	...a/	...a/		...	...	...a/	...a/
071	Mining of iron ores		...	...	...	...		...	...	...	...		...	...	...	...
072	Mining of non-ferrous metal ores		...	...	...	...		...	...	...	...		...	...	...	...
081	Quarrying of stone, sand and clay		...	...	8	21		...	...	2253	2401		...	...	286	233
089	Mining and quarrying n.e.c.		...	...	...	...		...	...	...	...		...	...	...	...
091	Support activities for petroleum		...	...	84	88		...	...	24299	20790		...	...	2472	2120
099	Support activities for other mining, quarrying		...	...	...	...		...	...	...	...		...	...	...	...
B	Mining and quarrying		...	...	99	116		...	...	46764	41842		...	...	13078	13127
351	Electric power generation, transmission		...	...	7b/	7b/		...	...	4826b/	4978b/		...	...	1298b/	1351b/
352	Manufacture of gas		...	...	...b/	...b/		...	...	...b/	...b/		...	...	...b/	...b/
353	Steam and air conditioning supply		...	...	...b/	...b/		...	...	...b/	...b/		...	...	...b/	...b/
D	Electricity, gas, steam & air conditioning		...	...	7	7		...	...	4826	4978		...	...	1298	1351
360	Water collection, treatment and supply		...	...	...	...		...	...	...	...		...	...	...	...
370	Sewerage		...	...	2	7		...	...	43	381		...	...	7	17
381	Waste collection		...	...	...	...		...	...	...	...		...	...	...	...
382	Waste treatment and disposal		...	...	-	1		...	...	-	387		...	...	-	52
383	Materials recovery		...	...	3	6		...	...	60	206		...	...	2	7
390	Remediation activities		...	...	5	8		...	...	128	475		...	...	7	38
E	Water supply;sewerage,waste management		...	...	10	22		...	...	231	1449		...	...	16	115

a/ 061 includes 062.
b/ 351 includes 352 and 353.

Qatar

ISIC Revision 4		Output at producers' prices					Value added at producers' prices				
		Note	(millions of Qatari Riyals)				Note	(millions of Qatari Riyals)			
ISIC	Industry		2012	2013	2014	2015		2012	2013	2014	2015
051	Mining of hard coal		...	...	...	...		...	...	...	...
052	Mining of lignite		...	...	...	...		...	...	...	...
061	Extraction of crude petroleum		...	...	420906a/	242625a/		...	...	384655a/	213207a/
062	Extraction of natural gas		...	...	...a/	...a/		...	...	...a/	...a/
071	Mining of iron ores		...	...	...	...		...	...	...	...
072	Mining of non-ferrous metal ores		...	...	...	...		...	...	...	...
081	Quarrying of stone, sand and clay		...	...	1335	813		...	...	876	437
089	Mining and quarrying n.e.c.		...	...	...	...		...	...	...	...
091	Support activities for petroleum		...	...	11590	9040		...	...	8251	7309
099	Support activities for other mining, quarrying		...	...	...	...		...	...	...	...
B	Mining and quarrying		...	...	433831	252478		...	...	393782	220953
351	Electric power generation, transmission		...	...	14325b/	15195b/		...	...	3228b/	3676b/
352	Manufacture of gas		...	...	...b/	...b/		...	...	...b/	...b/
353	Steam and air conditioning supply		...	...	...b/	...b/		...	...	...b/	...b/
D	Electricity, gas, steam & air conditioning		...	...	14325	15195		...	...	3228	3676
360	Water collection, treatment and supply		...	...	...	...		...	...	...	...
370	Sewerage		...	...	47	81		...	...	25	27
381	Waste collection		...	...	...	...		...	...	...	...
382	Waste treatment and disposal		...	...	-	210		...	...	-	-187
383	Materials recovery		...	...	9	41		...	...	4	16
390	Remediation activities		...	...	51	186		...	...	33	123
E	Water supply;sewerage,waste management		...	...	108	518		...	...	62	-21

a/ 061 includes 062.
b/ 351 includes 352 and 353.

Republic of Korea

ISIC Revision 4		Number of establishments					Number of employees					Wages and salaries paid to employees				
		Note	(number)				Note	(number)				Note	(billions of Korean Won)			
ISIC	Industry		2012	2013	2014	2015		2012	2013	2014	2015		2012	2013	2014	2015
051	Mining of hard coal		6a/	...	...	...		...	...	2364	2497		150a/	135	131	132
052	Mining of lignite		...a/	...	...	...		...	...	-	-		...a/	-	-	-
061	Extraction of crude petroleum		1b/	...	...	...		2949b/	2731b/	...	...		...	...	...	5
062	Extraction of natural gas		...b/	...	...	...		...b/	...b/	...	...		...	...	...	-
071	Mining of iron ores		7	...	...	...		441c/	405c/	212	144		14c/	7	8	6
072	Mining of non-ferrous metal ores		...	...	...	...		...c/	...c/	...	190		...c/	...	...	5
081	Quarrying of stone, sand and clay		312d/	...	...	...		12046d/	12191d/	6789	9225		236d/	248	266	362
089	Mining and quarrying n.e.c.		...d/	...	...	...		...d/	...d/	535	1172		...d/	14	17	28
091	Support activities for petroleum		2e/	...	...	...		144e/	545e/	...	...		...	...	...	...
099	Support activities for other mining, quarrying		...e/	...	...	...		...e/	...e/	...	...		...	...	...	...
B	Mining and quarrying		328	...	...	...		15580	15872	10169	13322		...	417	434	537
351	Electric power generation, transmission		...	...	...	...		114687f/	120729f/	...	42599		...	...	...	2993
352	Manufacture of gas		...	...	...	...		...f/	...f/	...	10596		...	...	...	685
353	Steam and air conditioning supply		...	...	...	...		...f/	...f/	...	3915		...	...	...	240
D	Electricity, gas, steam & air conditioning		...	...	...	...		114687	120729	...	57110		...	...	...	3918
360	Water collection, treatment and supply		...	...	...	...		15619	15865	...	17454		...	...	...	966
370	Sewerage		1485	...	...	...		15239	15427	15309	16195		503	526	520	680
381	Waste collection		1881	...	...	...		56852g/	61455g/	25332	26196		618	691	752	892
382	Waste treatment and disposal		1163	...	...	...		...g/	...g/	21119	21320		606	639	664	820
383	Materials recovery		1910	...	...	...		...g/	...g/	15721	15663		370	390	425	490
390	Remediation activities		85	...	...	...		995	1028	1045	1069		30	31	35	46
E	Water supply;sewerage,waste management		...	...	...	...		...	...	...	97897		...	...	...	3893

a/ 051 includes 052.
b/ 061 includes 062.
c/ 071 includes 072.
d/ 081 includes 089.
e/ 091 includes 099.
f/ 351 includes 352 and 353.
g/ 381 includes 382 and 383.

Republic of Korea

ISIC Revision 4		Output (valuation not defined)					Value added (valuation not defined)				
		Note	(billions of Korean Won)				Note	(billions of Korean Won)			
ISIC	Industry		2012	2013	2014	2015		2012	2013	2014	2015
051	Mining of hard coal		317a/	283	265	264		223a/	183	177	189
052	Mining of lignite		...a/	-	-	-		...a/	-	-	-
061	Extraction of crude petroleum		...	...	...	83		...	...	...	73
062	Extraction of natural gas		...	...	...	-		...	...	...	-
071	Mining of iron ores		82b/	72	67	38		73b/	67	61	35
072	Mining of non-ferrous metal ores		...b/	...	...	22		...b/	...	...	17
081	Quarrying of stone, sand and clay		2126c/	2290	2360	2492		1437c/	1522	1580	1687
089	Mining and quarrying n.e.c.		...c/	82	97	103		...c/	63	72	82
091	Support activities for petroleum		...	...	...	...		...	...	...	...
099	Support activities for other mining, quarrying		...	...	...	...		...	...	...	...
B	Mining and quarrying		...	3019	3013	3017		...	2105	2088	2097
351	Electric power generation, transmission		...	...	...	33672		...	...	...	23792
352	Manufacture of gas		...	...	...	6503		...	...	...	4031
353	Steam and air conditioning supply		...	...	...	2233		...	...	...	1137
D	Electricity, gas, steam & air conditioning		...	...	...	42408		...	...	...	28959
360	Water collection, treatment and supply		...	...	...	3802		...	...	...	1985
370	Sewerage		2965	2980	2975	2457		1046	1098	1245	1023
381	Waste collection		2277	2351	2586	2644		1054	1119	1278	1417
382	Waste treatment and disposal		3380	3615	3867	4217		1370	1421	1598	1844
383	Materials recovery		7601	7350	7208	1546		906	926	948	898
390	Remediation activities		164	160	172	163		56	62	64	73
E	Water supply;sewerage,waste management		...	...	...	14829		...	...	...	7241

a/ 051 includes 052.
b/ 071 includes 072.
c/ 081 includes 089.

Republic of Moldova

ISIC Revision 4		Number of enterprises					Number of employees					Wages and salaries paid to employees				
		Note	(number)				Note	(number)				Note	(thousands of Moldovan Lei)			
ISIC	Industry		2012	2013	2014	2015		2012	2013	2014	2015		2012	2013	2014	2015
051	Mining of hard coal		...	...	...	...		...	...	...	...		...	...	...	...
052	Mining of lignite		...	...	...	...		...	...	...	...		...	...	...	...
061	Extraction of crude petroleum		...	1	1	1		...	90	86	84		...	2564	3060	2855
062	Extraction of natural gas		...	...	...	...		...	...	...	...		...	...	...	...
071	Mining of iron ores		...	...	...	...		...	...	...	...		...	...	...	...
072	Mining of non-ferrous metal ores		...	...	...	...		...	...	...	...		...	...	...	...
081	Quarrying of stone, sand and clay		...	99	94	94		...	2005	2253	2046		...	95417	118369	112808
089	Mining and quarrying n.e.c.		...	...	...	...		...	...	...	...		...	...	...	...
091	Support activities for petroleum		...	...	...	...		...	267a/	...	...		...	10191a/	...	...
099	Support activities for other mining, quarrying		...	...	...	...		...	...a/	...	...		...	...a/	...	...
B	Mining and quarrying		...	100	95	95		...	2362	2344	2130		...	108172	121429	115662
351	Electric power generation, transmission		...	24	24	25		...	12230b/	11914b/	11665b/		...	942293b/	1036499b/	1147330b/
352	Manufacture of gas		...	19	19	18		...	...b/	...b/	...b/		...	...b/	...b/	...b/
353	Steam and air conditioning supply		...	18	18	15		...	...b/	...b/	...b/		...	...b/	...b/	...b/
D	Electricity, gas, steam & air conditioning		...	61	61	58		...	12230	11914	11665		...	942293	1036499	1147330
360	Water collection, treatment and supply		...	358	384	409		...	4547	4566	5047		...	201370	232452	261538
370	Sewerage		...	7	4	8		...	1152	1016	823		...	60763	64868	62050
381	Waste collection		...	3	4	6		...	1819c/	1632c/	1871c/		...	91178c/	89526c/	111736c/
382	Waste treatment and disposal		...	2	1	1		...	...c/	...c/	...c/		...	...c/	...c/	...c/
383	Materials recovery		...	82	78	25		...	...c/	...c/	...c/		...	...c/	...c/	...c/
390	Remediation activities		...	...	...	...		...	...	...	...		...	...	...	...
E	Water supply;sewerage,waste management		...	452	471	449		...	7517	7413	7741		...	353311	386846	435323

a/ 091 includes 099.
b/ 351 includes 352 and 353.
c/ 381 includes 382 and 383.

Republic of Moldova

ISIC Revision 4		Output (valuation not defined)					Value added (valuation not defined)				
			(millions of Moldovan Lei)					(millions of Moldovan Lei)			
ISIC	Industry	Note	2012	2013	2014	2015	Note	2012	2013	2014	2015
051	Mining of hard coal		...	...	...	...		...	...	...	...
052	Mining of lignite		...	...	...	...		...	...	...	...
061	Extraction of crude petroleum		...	...	...	...		...	...	...	...
062	Extraction of natural gas		...	...	...	...		...	...	...	...
071	Mining of iron ores		...	...	...	...		...	...	...	...
072	Mining of non-ferrous metal ores		...	...	...	...		...	...	...	...
081	Quarrying of stone, sand and clay		...	683.6	760.1	734.6		...	286.1	370.1	382.0
089	Mining and quarrying n.e.c.		...	...	...	...		...	...	...	...
091	Support activities for petroleum		...	...	...	...		...	...	...	...
099	Support activities for other mining, quarrying		...	...	...	...		...	...	...	...
B	Mining and quarrying		...	730.4	799.5	760.9		...	305.2	389.3	395.6
351	Electric power generation, transmission		...	3042.5	3130.9	3879.9		...	1133.4	991.5	1294.9
352	Manufacture of gas		...	793.3	818.4	836.3		...	...	...	...
353	Steam and air conditioning supply		...	891.3	897.8	287.9		...	331.9	284.6	96.1
D	Electricity, gas, steam & air conditioning		...	4727.1	4847.1	5004.0		...	1465.4	1276.1	1391.0
360	Water collection, treatment and supply		...	969.5	977.6	1032.6		...	...	...	...
370	Sewerage		...	1.4	28.9	31.1		...	...	...	...
381	Waste collection		...	2.1	2.1	2.4		...	...	...	...
382	Waste treatment and disposal		...	...	...	...		...	...	...	...
383	Materials recovery		...	1265.5	1440.1	1117.2		...	...	...	...
390	Remediation activities		...	...	...	...		...	...	...	...
E	Water supply;sewerage,waste management		...	2238.5	2449.1	2183.4		...	...	...	...

Romania

ISIC Revision 4		Number of establishments					Number of employees					Wages and salaries paid to employees				
			(number)					(number)					(millions of Romanian Lei)			
ISIC	Industry	Note	2012	2013	2014	2015	Note	2012	2013	2014	2015	Note	2012	2013	2014	2015
05	Mining of coal and lignite		75	57	55	48		11482	5754	2207	1838		134	233	...	...
051	Mining of hard coal		27	18	15	12		7719	5373	1867	1571		...	...	...	...
052	Mining of lignite		48	39	40	36		3763	381	340	267		...	...	7	...
06	Extraction of crude petroleum & natural gas		76	77	79	77		26547	25677	25243	23482		...	...	...	...
061	Extraction of crude petroleum		52	56	55	54		20528	19097	18615	16952		...	...	...	...
062	Extraction of natural gas		24	21	24	23		6019	6580	6628	6530		...	...	...	...
07	Mining of metal ores		45	42	46	44		2672	2584	2465	2488		...	...	93	...
071	Mining of iron ores		7	2	2	3		71	36	27	30		...	...	...	...
072	Mining of non-ferrous metal ores		38	40	44	41		2601	2548	2438	2458		...	...	...	...
08	Other mining and quarrying		1002	972	996	976		10420	9830	9523	10387		214	229	217	281
081	Quarrying of stone, sand and clay		944	914	929	908		7934	7431	7210	8179		...	...	...	...
089	Mining and quarrying n.e.c.		58	58	67	68		2486	2399	2313	2208		...	...	...	...
09	Mining support service activities		138	137	149	148		6904	7800	7584	6951		301	428	434	413
091	Support activities for petroleum		114	113	128	126		6622	7508	7371	6721		295	418	429	408
099	Support activities for other mining, quarrying		24	24	21	22		282	292	213	230		6	10	6	6
B	Mining and quarrying		1336a/	1285a/	1325a/	1293		58025	51645	47022	45146		2814	2790	3000	3035
35	Electricity, gas, steam & air conditioning		1192	1467	1643	1584		79304	76916	75426	71780		3772	3645	3727	3668
351	Electric power generation, transmission		984	1254	1426	1358		53660	55517	55350	51596		2765	2753	2874	2754
352	Manufacture of gas		95	100	98	108		10002	7158	6766	7107		514	400	400	435
353	Steam and air conditioning supply		113	113	119	118		15642	14241	13310	13077		493	492	453	479
D	Electricity, gas, steam & air conditioning		1192	1467	1643	1584		79304	76916	75426	71780		3772	3645	3727	3668
36	Water collection, treatment and supply		431	465	495	508		36002	36648	36450	35868		969	1036	1089	1160
37	Sewerage		194	211	198	207		1460	1437	1495	1424		40	39	49	42
38	Waste collection,treatment,disposal activities		2585	2674	2784	2668		49072	48522	46298	45747		873	892	863	1022
381	Waste collection		924	985	1043	1024		32266	32101	31026	31258		592	599	579	688
382	Waste treatment and disposal		109	129	148	157		1813	2278	2277	2380		41	53	57	67
383	Materials recovery		1552	1560	1593	1487		14993	14143	12995	12109		240	240	226	267
39	Remediation activities		63	63	66	57		601	569	531	540		15	13	9	11
E	Water supply;sewerage,waste management		3273	3413	3543	3440		87135	87176	84774	83579		1897	1980	2010	2235

a/ Sum of available data.

Romania

ISIC Revision 4		Output at factor values a/					Value added at factor values				
			(millions of Romanian Lei)					(millions of Romanian Lei)			
ISIC	Industry	Note	2012	2013	2014	2015	Note	2012	2013	2014	2015
05	Mining of coal and lignite		1050	451	146	91		307	418	...	...
051	Mining of hard coal		495	359	100	62		...	...	...	...
052	Mining of lignite		555	92	46	29		...	...	17	...
06	Extraction of crude petroleum & natural gas		23641	22323	21786	18876		...	...	...	...
061	Extraction of crude petroleum		19711	18233	17038	14531		...	...	...	...
062	Extraction of natural gas		3930	4090	4748	4345		...	...	...	...
07	Mining of metal ores		337	365	365	360		...	...	176	...
071	Mining of iron ores		3	2	2	4		...	...	...	...
072	Mining of non-ferrous metal ores		334	363	363	357		...	...	...	...
08	Other mining and quarrying		2040	1851	1957	2494		582	853	676	855
081	Quarrying of stone, sand and clay		1689	1400	1578	2090		...	...	...	...
089	Mining and quarrying n.e.c.		351	451	379	404		...	...	...	...
09	Mining support service activities		2799	3573	3396	2875		768	927	1194	735
091	Support activities for petroleum		2737	3519	3366	2834		755	887	1182	724
099	Support activities for other mining, quarrying		62	54	30	42		13	40	12	11
B	Mining and quarrying		29867	28563	27650	24697		16069	15927	15003	11509
35	Electricity, gas, steam & air conditioning		59718	55996	55120	58567		15098	20341	17519	14545
351	Electric power generation, transmission		42781	39905	39148	41591		12565	17417	14836	12302
352	Manufacture of gas		14315	13421	13526	14664		1864	2262	2132	2269
353	Steam and air conditioning supply		2622	2670	2447	2312		669	663	551	-26
D	Electricity, gas, steam & air conditioning		59718	55996	55120	58567		15098	20341	17519	14545
36	Water collection, treatment and supply		3167	3403	3582	3754		2068	2057	2263	2386
37	Sewerage		229	208	278	302		91	79	118	75
38	Waste collection,treatment,disposal activities		13042	11448	10225	8473		2055	2131	2001	1995
381	Waste collection		2679	2839	2903	2803		1171	1273	1122	1289
382	Waste treatment and disposal		326	461	418	451		121	217	213	200
383	Materials recovery		10037	8148	6904	5218		763	641	666	505
39	Remediation activities		235	124	127	148		98	47	53	70
E	Water supply;sewerage,waste management		16673	15183	14212	12677		4312	4314	4435	4526

a/ Output refers to turnover.

Russian Federation

ISIC Revision 3		Number of establishments					Number of employees					Wages and salaries paid to employees a/				
			(number)					(thousands)					(millions of Russian Roubles)			
ISIC	Industry	Note	2012	2013	2014	2015	Note	2012	2013	2014	2015	Note	2012	2013	2014	2015
101	Mining and agglomeration of hard coal		494	516	509	463		146	142	132	126		64502	66734	65873	68874
102	Mining and agglomeration of lignite		90	108	107	98		20	19	17	17		7010	7165	7556	8285
103	Extraction and agglomeration of peat		90	204	195	193		3	3	3	3		542	558	564	578
111	Extraction of crude oil and natural gas		820	737	745	758		185	179	184	190		166028	175025	195357	215241
112	Oil, gas service activities		1435	1506	1545	1611		244	255	270	286		172939	189261	215920	239128
120	Mining of uranium and thorium ores	b/	...	...	...	...	b/	...	...	...	...	b/	...	...	...	...
131	Iron ores		32	95	103	82		63	62	62	59		25632	27409	29023	29424
132	Non-ferrous metal ores,excl.uranium,thorium		317	1151	1198	1478		133	132	128	130		74712	79405	81681	94387
141	Quarrying of stone, sand and clay		3928	4153	4436	5019		44	43	43	41		12805	13668	14279	13474
142	Mining and quarrying n.e.c.		335	311	343	423		50	54	54	51		13071	15252	15975	15517
C	Total mining and quarrying		7547	8786	9187	10125		886	938	933	942		537241	609506	660034	720161
401	Electricity production, collection, distribution		5327	5463	5710	5627		713	717	714	711		340772	373447	402385	420902
402	Gas production and distribution		703	581	586	598		161	166	162	164		54496	60723	63309	67864
403	Steam and hot water supply		8481	8458	8500	8973		682	676	664	648		186615	202151	213869	222596
410	Water collection, purification, distribution		4312	4371	4401	4584		277	271	274	272		65642	70994	78000	82648
E	Total electricity, gas and water supply		18823	18873	19197	19782		1833	1829	1814	1795		647532	707320	757563	794010

a/ Data were computed from the reported monthly average wages.

b/ Data suppressed due to confidentiality rules.

Russian Federation

ISIC Revision 3		Output at basic prices					Value added at basic prices				
			(billions of Russian Roubles)					(billions of Russian Roubles)			
ISIC	Industry	Note	2012	2013	2014	2015	Note	2012	2013	2014	2015
101	Mining and agglomeration of hard coal		631.6	577.3	602.7	741.6		255.9	209.0	217.2	272.8
102	Mining and agglomeration of lignite		45.5	49.2	51.7	58.7		23.6	24.9	27.7	31.2
103	Extraction and agglomeration of peat		2.5	2.2	2.4	3.4		1.1	0.9	1.0	1.4
111	Extraction of crude oil and natural gas		6078.1	6315.5	6702.5	7591.3		4622.0	4696.9	5039.7	5815.4
112	Oil, gas service activities		935.5	1014.9	1195.2	1363.6		318.1	350.4	411.4	455.4
120	Mining of uranium and thorium ores		...	...	...	...		...	...	...	...
131	Iron ores		312.0	323.8	295.2	299.6		175.4	174.9	147.3	150.8
132	Non-ferrous metal ores,excl.uranium,thorium		409.6	132.8	174.9	193.7		206.5	59.9	90.6	106.5
141	Quarrying of stone, sand and clay		412.0a/	473.1	531.9	629.5		220.8a/	272.7	304.8	382.7
142	Mining and quarrying n.e.c.		...a/	...	...	...		...a/	...	...	...
C	Total mining and quarrying		8831.1	8892.4	9559.6	10884.4		5826.1	5791.6	6241.5	7217.5
401	Electricity production, collection, distribution		3762.6	3885.0	4152.9	4237.9		1296.9	1413.1	1452.7	1575.2
402	Gas production and distribution		288.9	1236.2	1414.9	1367.2		142.0	107.1	107.3	123.4
403	Steam and hot water supply		993.6	1034.9	1137.0	1327.1		247.5	257.0	280.9	358.4
410	Water collection, purification, distribution		270.2	278.0	286.9	314.5		137.8	144.2	151.7	165.7
E	Total electricity, gas and water supply		5315.4	6434.1	6991.8	7246.8		1824.2	1921.5	1992.5	2222.6

a/ 141 includes 142.

Saudi Arabia

ISIC Revision 4		Number of establishments					Number of employees					Wages and salaries paid to employees				
			(number)					(number)					(millions of Saudi Riyals)			
ISIC	Industry	Note	2012	2013	2014	2015	Note	2012	2013	2014	2015	Note	2012	2013	2014	2015
05	Mining of coal and lignite		...	...	...	5		...	...	...	12		...	...	...	-
06	Extraction of crude petroleum & natural gas		...	...	...	69		...	...	...	60980		...	...	...	31756
07	Mining of metal ores		...	...	...	48		...	...	...	3680		...	...	...	182
08	Other mining and quarrying		...	...	...	374		...	...	...	9832		...	...	...	330
09	Mining support service activities		...	...	...	143		...	...	...	20182		...	...	...	...
B	Mining and quarrying		...	...	...	639		...	...	...	94686		...	...	...	32274
35	Electricity, gas, steam & air conditioning		...	...	...	754		...	...	...	68181		...	...	...	8239
D	Electricity, gas, steam & air conditioning		...	...	...	754		...	...	...	68181		...	...	...	8239
36	Water collection, treatment and supply		...	...	...	1918		...	...	...	19995		...	...	...	507
37	Sewerage		...	...	...	132		...	...	...	10116		...	...	...	397
38	Waste collection,treatment,disposal activities		...	...	...	278		...	...	...	9859		...	...	...	375
39	Remediation activities		...	...	...	18		...	...	...	258		...	...	...	10
E	Water supply;sewerage,waste management		...	...	...	2346		...	...	...	40228		...	...	...	1290

Saudi Arabia

ISIC Revision 4		Output					Value added at producers' prices				
		Note	(millions of Saudi Riyals)				Note	(millions of Saudi Riyals)			
ISIC	Industry		2012	2013	2014	2015		2012	2013	2014	2015
05	Mining of coal and lignite		...	...	...	...		...	...	...	1
06	Extraction of crude petroleum & natural gas		...	...	...	...		...	...	...	611739
07	Mining of metal ores		...	...	...	...		...	...	...	2017
08	Other mining and quarrying		...	...	...	...		...	...	...	6221
09	Mining support service activities		...	...	...	...		...	...	...	2268
B	Mining and quarrying		...	...	...	...		...	...	...	622246
35	Electricity, gas, steam & air conditioning		...	...	...	...		...	...	...	31453
D	Electricity, gas, steam & air conditioning		...	...	...	...		...	...	...	31453
36	Water collection, treatment and supply		...	...	...	...		...	...	...	2576
37	Sewerage		...	...	...	...		...	...	...	1226
38	Waste collection,treatment,disposal activities		...	...	...	...		...	...	...	789
39	Remediation activities		...	...	...	...		...	...	...	19
E	Water supply;sewerage,waste management		...	...	...	...		...	...	...	4610

Serbia

ISIC Revision 4		Output at basic prices					Value added at basic prices				
		Note	(millions of Serbian Dinars)				Note	(millions of Serbian Dinars)			
ISIC	Industry		2012	2013	2014	2015		2012	2013	2014	2015
05	Mining of coal and lignite		52119	52408	42020	45609		33274	35310	24867	22390
06	Extraction of crude petroleum & natural gas		16721	15407	14052	9388		7676	7222	5066	1956
07	Mining of metal ores		22659	20236	17614	17556		10148	9046	5644	7010
08	Other mining and quarrying		6918	8527	9378	10682		1863	2013	2361	2488
09	Mining support service activities		11017	20392	21853	22726		5020	5244	5449	4785
B	Mining and quarrying		109435	116971	104917	105960		57980	58835	43386	38629
35	Electricity, gas, steam & air conditioning		493825	586932	515565	659661		103711	138806	112643	149745
D	Electricity, gas, steam & air conditioning		493825	586932	515565	659661		103711	138806	112643	149745
36	Water collection, treatment and supply		37633	37529	39157	39646		21908	24144	24363	23506
37	Sewerage		202	240	252	256		91	105	106	102
38	Waste collection,treatment,disposal activities		63600	59123	56971	53607		21338	22015	23172	22450
39	Remediation activities		1074	1198	1179	1201		232	248	268	338
E	Water supply;sewerage,waste management		102509	98090	97558	94710		43570	46512	47908	46396

Slovakia

ISIC Revision 4		Number of enterprises					Number of employees					Wages and salaries paid to employees				
			(number)					(number)					(millions of Euros)			
ISIC	Industry	Note	2012	2013	2014	2015	Note	2012	2013	2014	2015	Note	2012	2013	2014	2015
051	Mining of hard coal		...	...	-	-		...	...	-	-		...	...	-	-
052	Mining of lignite		...	...	...	...		...	...	...	...		...	...	...	...
061	Extraction of crude petroleum		-	-	-	-		-	-	-	-		-	-	-	-
062	Extraction of natural gas		...	...	...	...		...	...	...	...		...	...	...	...
071	Mining of iron ores		-	-	-	-		-	-	-	-		-	-	-	-
072	Mining of non-ferrous metal ores		...	...	...	...		...	...	...	...		...	...	...	...
081	Quarrying of stone, sand and clay		81	106	114	103		1682	1820	1524	1524		21	21	18	20
089	Mining and quarrying n.e.c.		29	31	51	65		276	311	314	319		3	3	3	3
091	Support activities for petroleum		...	...	...	...		...	...	...	...		...	...	...	...
099	Support activities for other mining, quarrying		7	11	10	11		254	258	244	300		7	7	6	6
B	Mining and quarrying		125	157	182	187		7267	7332	6896	6743		91	91	89	90
351	Electric power generation, transmission		173	223	350	296		9550	9461	10197	10339		215	217	239	240
352	Manufacture of gas		8	12	11	10		2581	2442	2336	2233		62	59	55	53
353	Steam and air conditioning supply		177	195	170	145		6290	5790	5174	5100		81	76	73	73
D	Electricity, gas, steam & air conditioning		358	430	531	451		18421	17693	17707	17672		358	353	367	365
360	Water collection, treatment and supply		47	56	51	48		11623	11503	11342	11194		126	128	127	128
370	Sewerage		96	112	120	143		297	386	446	383		3	4	5	4
381	Waste collection		201	240	266	290		4280	4814	4232	4332		41	45	43	44
382	Waste treatment and disposal		123	155	204	275		2654	2797	2857	2968		36	36	38	42
383	Materials recovery		445	433	394	333		1760	1577	1551	940		17	23	15	8
390	Remediation activities		17	19	24	23		225	196	194	111		2	2	2	1
E	Water supply;sewerage,waste management		929	1015	1059	1112		20839	21273	20622	19928		226	238	230	227

Slovakia

ISIC Revision 4		Output (valuation not defined)					Value added at factor values				
			(millions of Euros)					(millions of Euros)			
ISIC	Industry	Note	2012	2013	2014	2015	Note	2012	2013	2014	2015
051	Mining of hard coal		...	...	-	-		...	...	-	-
052	Mining of lignite		...	...	...	...		...	...	...	...
061	Extraction of crude petroleum		-	-	-	-		-	-	-	-
062	Extraction of natural gas		...	...	...	...		...	...	...	...
071	Mining of iron ores		-	-	-	-		-	-	-	-
072	Mining of non-ferrous metal ores		...	...	...	...		...	...	...	...
081	Quarrying of stone, sand and clay		108	136	127	154		34	35	42	49
089	Mining and quarrying n.e.c.		11	14	17	22		4	5	5	7
091	Support activities for petroleum		...	...	...	...		...	...	...	...
099	Support activities for other mining, quarrying		14	15	17	20		10	10	10	12
B	Mining and quarrying		498	528	520	563		314	318	318	346
351	Electric power generation, transmission		8219	7698	6538	6726		1729	1548	1505	1431
352	Manufacture of gas		2901	2813	2235	2094		536	153	260	315
353	Steam and air conditioning supply		1060	1049	900	860		234	245	227	220
D	Electricity, gas, steam & air conditioning		12180	11560	9673	9680		2499	1945	1992	1966
360	Water collection, treatment and supply		482	486	491	488		269	266	268	303
370	Sewerage		20	24	47	25		9	10	20	13
381	Waste collection		186	264	190	220		96	136	92	103
382	Waste treatment and disposal		116	127	149	209		26	31	27	53
383	Materials recovery		131	172	81	59		51	44	26	12
390	Remediation activities		8	9	8	5		3	4	3	2
E	Water supply;sewerage,waste management		942	1080	965	1007		453	491	435	485

Slovenia

ISIC Revision 4		Number of enterprises					Number of employees					Wages and salaries paid to employees				
			(number)					(number)					(millions of Euros)			
ISIC	Industry	Note	2012	2013	2014	2015	Note	2012	2013	2014	2015	Note	2012	2013	2014	2015
051	Mining of hard coal		-	-	-	-		-	-	-	-		-	-	-	-
052	Mining of lignite		1	1	1	1		...	...	...	...		...	...	...	...
061	Extraction of crude petroleum		-	-	-	-		-	-	-	-		-	-	-	-
062	Extraction of natural gas		1	1	1	1		-	-	...	...		...	...	...	...
071	Mining of iron ores		-	-	-	-		-	-	-	-		-	-	-	-
072	Mining of non-ferrous metal ores		-	-	-	-		-	-	-	-		-	-	-	-
081	Quarrying of stone, sand and clay		88	86	84	80		836	777	781	854		17	17	17	20
089	Mining and quarrying n.e.c.		10	12	9	11		174	168	...	...		...	...	3	...
091	Support activities for petroleum		3	4	2	2		...	68	...	...		...	2a/	...	...
099	Support activities for other mining, quarrying		3	1	2	2		7	...	...	...		...	...a/	...	...
B	Mining and quarrying		106	105	99	97		2435	2359	2370	2380		68	65	68	65
351	Electric power generation, transmission		1248	1464	1500	1461		6870	6848	6660	6589		211	220	216	215
352	Manufacture of gas		10	9	10	10		138	137	138	139		4	4	4	5
353	Steam and air conditioning supply		47	53	60	59		772	774	1053	1055		19	19	27	29
D	Electricity, gas, steam & air conditioning		1305	1526	1570	1530		7781	7759	7850	7783		234	244	248	250
360	Water collection, treatment and supply		70	71	75	74		3819	3908	3995	4008		79	81	85	86
370	Sewerage		34	40	42	42		311	318	351	359		6	7	7	8
381	Waste collection		119	119	130	126		3000	2993	3067	3071		60	60	61	63
382	Waste treatment and disposal		32	36	38	39		262	266	...	292		5	5	6	6
383	Materials recovery		110	103	97	91		1898	1841	...	1573		37	34	31	30
390	Remediation activities		27	32	31	29		242	360	448	438		4	7	8	8
E	Water supply;sewerage,waste management		392	401	413	401		9532	9687	9811	9741		191	194	199	201

a/ 091 includes 099.

Slovenia

ISIC Revision 4		Output at factor values					Value added at factor values				
			(millions of Euros)					(millions of Euros)			
ISIC	Industry	Note	2012	2013	2014	2015	Note	2012	2013	2014	2015
051	Mining of hard coal		-	-	-	-		-	-	-	-
052	Mining of lignite		...	...	...	...		...	...	...	...
061	Extraction of crude petroleum		-	-	-	-		-	-	-	-
062	Extraction of natural gas		...	...	...	...		...	...	...	...
071	Mining of iron ores		-	-	-	-		-	-	-	-
072	Mining of non-ferrous metal ores		-	-	-	-		-	-	-	-
081	Quarrying of stone, sand and clay		117	127	145	151		39	40	49	50
089	Mining and quarrying n.e.c.		...	...	9	...		...	...	5	...
091	Support activities for petroleum		...	5a/	...	...		...	1a/	...	...
099	Support activities for other mining, quarrying		...	...a/	...	...		...	...a/	...	...
B	Mining and quarrying		251	252	262	268		109	104	109	111
351	Electric power generation, transmission		2554	2575	2432	2206		706	740	729	658
352	Manufacture of gas		40	41	41	44		16	15	16	20
353	Steam and air conditioning supply		206	196	188	202		43	49	55	67
D	Electricity, gas, steam & air conditioning		2801	2812	2661	2452		765	804	799	745
360	Water collection, treatment and supply		315	340	384	378		116	125	135	131
370	Sewerage		24	25	27	30		10	11	11	12
381	Waste collection		249	263	278	265		91	98	99	98
382	Waste treatment and disposal		32	31	34	32		10	11	12	10
383	Materials recovery		391	353	325	292		65	53	50	50
390	Remediation activities		14	30	32	28		5	11	14	14
E	Water supply;sewerage,waste management		1025	1042	1079	1026		299	308	322	316

a/ 091 includes 099.

South Africa

ISIC Revision 3		Number of establishments					Number of employees a/					Wages and salaries paid to employees a/			
		Note	(number)				Note	(number)				Note	(millions of South African Rand)			
ISIC	Industry		2012	2013	2014	2015		2012	2013	2014	2015		2012 b/	2013 b/	2014 b/	2015 b/
101	Mining and agglomeration of hard coal		293c/	315c/	345c/	...		...	...	...	...		...	...	...	...
102	Mining and agglomeration of lignite		...c/	...c/	...c/	...		...	...	...	...		...	...	...	...
103	Extraction and agglomeration of peat		...c/	...c/	...c/	...		...	...	...	...		...	...	...	...
111	Extraction of crude oil and natural gas		55d/	59d/	64d/	...		...	...	...	...		...	...	...	...
112	Oil, gas service activities		...d/	...d/	...d/	...		...	...	...	...		...	...	...	...
120	Mining of uranium and thorium ores		124	119	125	...		...	...	...	...		...	...	...	...
131	Iron ores		30	27	29	...		...	...	...	...		...	...	...	...
132	Non-ferrous metal ores,excl.uranium,thorium		412	344	349	...		...	...	...	...		...	...	...	...
141	Quarrying of stone, sand and clay		581	549	569	...		...	...	...	...		...	...	...	...
142	Mining and quarrying n.e.c.		1626	1584	1698	...		...	...	...	...		...	...	...	...
C	Total mining and quarrying		3121	2997	3179	...		524632	510099	495568	478868		93630	100753	102146	114008
401	Electricity production, collection, distribution		516	572	645	...		...	...	...	...		...	...	...	...
402	Gas production and distribution		144	160	174	...		...	...	...	...		...	...	...	...
403	Steam and hot water supply		5	5	5	...		...	...	...	...		...	...	...	...
410	Water collection, purification, distribution		327	344	377	...		...	...	...	...		...	...	...	...
E	Total electricity, gas and water supply		992	1081	1201	...		...	...	...	...		...	...	...	...

a/ Data were derived from the national publication Facts and Figures 2016.

b/ Data refer to remuneration paid to employees.

c/ 101 includes 102 and 103.

d/ 111 includes 112.

South Africa

ISIC Revision 3		Output (valuation not defined) a/					Value added (valuation not defined) a/				
		Note	(millions of South African Rand)				Note	(millions of South African Rand)			
ISIC	Industry		2012b/	2013b/	2014b/	2015b/		2012c/	2013c/	2014c/	2015c/
10	Mining of coal and lignite; extraction of peat		...	...	...	...		...	...	...	...
11	Extraction of crude oil and natural gas, etc.		...	...	...	...		...	...	...	...
12	Mining of uranium and thorium ores		...	...	...	...		...	...	...	...
13	Mining of metal ores		...	...	...	...		...	...	...	...
14	Other mining and quarrying		...	...	...	...		...	...	...	...
C	Total mining and quarrying		364117	397678	396278	387647		267344	288085	286163	284012
40	Electricity, gas, steam and hot water supply		...	...	...	...		...	...	...	...
41	Water collection, purification, distribution		...	...	...	...		...	...	...	...
E	Total electricity, gas and water supply		...	...	...	...		...	...	...	...

a/ Data were derived from the national publication Facts and Figures 2016.

b/ Output refers to the total primary minerals sales.

c/ Value added is defined as total value added.

Spain

ISIC Revision 4		Number of enterprises					Number of employees					Wages and salaries paid to employees				
			(number)					(thousands)					(millions of Euros)			
ISIC	Industry	Note	2012	2013	2014	2015	Note	2012	2013	2014	2015	Note	2012	2013	2014	2015
051	Mining of hard coal		32	56	25	81a/		4.8	4.0	3.7	1.7a/		151	135	125	55a/
052	Mining of lignite		-	-	-	...a/		-	-	-	...a/		-	-	-	...a/
061	Extraction of crude petroleum		4	4	18b/	19b/		0.3	0.3	0.3b/	0.4b/		25	32	37b/	48b/
062	Extraction of natural gas		...	...	...b/	...b/		...	...	...b/	...b/		...	...	...b/	...b/
071	Mining of iron ores		19	...	9	22		0.1	...	0.1	0.1		2	...	2	2
072	Mining of non-ferrous metal ores		40	12	15	48		1.4	1.1	1.2	1.5		52	41	37	55
081	Quarrying of stone, sand and clay		2074	1729	1720	1839		13.9	11.8	11.1	11.1		362	295	279	285
089	Mining and quarrying n.e.c.		145	102	108	157		2.8	2.7	2.8	2.8		107	104	106	108
091	Support activities for petroleum		38	48	17c/	46		0.2	0.2	0.2c/	0.2		8	8	7c/	9
099	Support activities for other mining, quarrying		...	4	...c/	7		...	-	...c/	-		...	1	...c/	1
B	Mining and quarrying		2357	1967	1911	2219		23.5	20.3	19.4	17.7		708	619	592	563
351	Electric power generation, transmission		13722	13650	14022	13804		32.9	30.8	30.4	30.9		1859	1863	1815	1768
352	Manufacture of gas		77	58	68	63		4.8	5.0	4.6	4.1		320	307	306	274
353	Steam and air conditioning supply		186	160	153	178		0.8	0.7	0.8	1.0		14	12	15	18
D	Electricity, gas, steam & air conditioning		13986	13867	14244	14044		38.4	36.6	35.8	36.0		2193	2183	2136	2059
360	Water collection, treatment and supply		2794	2919	2732	3075		39.8	39.6	37.6	42.6		1236	1204	1150	1313
370	Sewerage		461	480	477	511		4.2	4.6	4.9	6.0		111	121	137	146
381	Waste collection		1592	1598	1420	1570		33.4	59.0	69.1	69.9		809	1458	1658	1732
382	Waste treatment and disposal		431	356	414	513		28.3	36.0	25.1	26.5		752	943	668	694
383	Materials recovery		568	555	536	498		8.7	8.7	8.6	8.5		209	214	205	201
390	Remediation activities		288	398	148	285		2.5	1.7	1.3	1.2		59	41	34	31
E	Water supply;sewerage,waste management		6134	6306	5726	6452		116.8	149.5	146.7	154.7		3176	3981	3852	4116

a/ 051 includes 052.
b/ 061 includes 062.
c/ 091 includes 099.

Spain

ISIC Revision 4		Output (valuation not defined)					Value added at factor values				
			(millions of Euros)					(millions of Euros)			
ISIC	Industry	Note	2012	2013	2014	2015	Note	2012	2013	2014	2015
051	Mining of hard coal		552	415	422	268a/		175	116	151	91a/
052	Mining of lignite		-	-	-	...a/		-	-	-	...a/
061	Extraction of crude petroleum		77	92	108b/	128b/		43	55	45b/	53b/
062	Extraction of natural gas		...	...	...b/	...b/		...	...	...b/	...b/
071	Mining of iron ores		7	...	7	7		2	...	2	2
072	Mining of non-ferrous metal ores		823	411	368	476		476	189	155	166
081	Quarrying of stone, sand and clay		1865	1556	1558	1641		697	606	570	575
089	Mining and quarrying n.e.c.		871	865	814	850		377	361	307	325
091	Support activities for petroleum		131	246	178c/	102		67	153	135c/	54
099	Support activities for other mining, quarrying		...	7	...c/	3		...	2	...c/	2
B	Mining and quarrying		4341	3614	3455	3475		1848	1496	1366	1267
351	Electric power generation, transmission		51723	63022	64254	52805		19440	17920	17016	18235
352	Manufacture of gas		9777	10621	10208	7933		3286	2471	2542	2362
353	Steam and air conditioning supply		82	76	90	115		37	36	35	46
D	Electricity, gas, steam & air conditioning		61581	73719	74552	60853		22763	20426	19593	20643
360	Water collection, treatment and supply		6931	7213	7356	7821		3289	3527	3460	3852
370	Sewerage		476	604	633	616		242	261	299	288
381	Waste collection		2530	4181	4246	4256		1441	3023	3015	2889
382	Waste treatment and disposal		3326	3960	3619	3545		1645	2064	1753	1808
383	Materials recovery		1676	1751	1648	1960		426	466	422	422
390	Remediation activities		208	159	141	156		104	73	63	65
E	Water supply;sewerage,waste management		15148	17867	17643	18353		7147	9413	9010	9324

a/ 051 includes 052.
b/ 061 includes 062.
c/ 091 includes 099.

Sri Lanka

ISIC Revision 4		Number of establishments					Number of employees					Wages and salaries paid to employees				
			(number)					(number)					(millions of Sri Lankan Rupees)			
ISIC	Industry	Note	2012	2013	2014	2015	Note	2012	2013	2014	2015	Note	2012	2013	2014	2015
05	Mining of coal and lignite		...	...	...	...		...	...	...	...		...	...	...	...
051	Mining of hard coal		...	...	...	...		...	...	...	...		...	...	...	...
052	Mining of lignite		...	...	...	...		...	...	...	...		...	...	...	...
06	Extraction of crude petroleum & natural gas		...	...	...	...		...	...	...	...		...	...	...	...
061	Extraction of crude petroleum		...	...	...	...		...	...	...	...		...	...	...	...
062	Extraction of natural gas		...	...	...	...		...	...	...	...		...	...	...	...
07	Mining of metal ores		...	...	...	...		...	...	...	...		...	...	...	...
071	Mining of iron ores		...	...	...	...		...	...	...	...		...	...	...	...
072	Mining of non-ferrous metal ores		...	...	...	...		...	...	...	...		...	...	...	...
08	Other mining and quarrying		...	124	79	...		...	23413	5994	...		...	1565.9	587.3	...
081	Quarrying of stone, sand and clay		...	50	31	...		...	19331	1056	...		...	1288.6	25.6	...
089	Mining and quarrying n.e.c.		...	74	48	...		...	4082	4938	...		...	277.3	561.7	...
09	Mining support service activities		...	...	...	...		...	...	...	...		...	...	...	...
091	Support activities for petroleum		...	...	...	...		...	...	...	...		...	...	...	...
099	Support activities for other mining, quarrying		...	...	...	...		...	...	...	...		...	...	...	...
B	Mining and quarrying	a/	...	124	79	...	a/	...	23413	5994	...	a/	...	1565.9	587.3	...
35	Electricity, gas, steam & air conditioning		...	10	37	...		...	16775	18690	...		...	12128.6	12682.8	...
351	Electric power generation, transmission		...	4	4	...		...	16407	16336	...		...	12082.7	11837.3	...
352	Manufacture of gas		...	...	...	...		...	...	...	...		...	...	...	...
353	Steam and air conditioning supply		...	6	33	...		...	368	2354	...		...	45.9	845.5	...
D	Electricity, gas, steam & air conditioning	a/	...	10	37	...	a/	...	16775	18690	...	a/	...	12128.6	12682.8	...
36	Water collection, treatment and supply		...	...	1	...		...	...	10424	...		...	...	7405.7	...
37	Sewerage		...	...	...	...		...	...	...	...		...	...	...	...
38	Waste collection,treatment,disposal activities		...	2	12	...		...	648	364	...		...	81.5	113.0	...
381	Waste collection		...	...	...	...		...	...	...	...		...	...	...	...
382	Waste treatment and disposal		...	2	...	...		...	648	...	...		...	81.5	...	...
383	Materials recovery		...	...	12	...		...	...	364	...		...	...	113.0	...
39	Remediation activities		...	3	...	...		...	179	...	...		...	84.8	...	...
E	Water supply;sewerage,waste management		...	...	...	...		...	...	...	...		...	...	...	...

a/ Sum of available data.

Sri Lanka

ISIC Revision 4		Output at basic prices					Value added at basic prices				
			(millions of Sri Lankan Rupees)					(millions of Sri Lankan Rupees)			
ISIC	Industry	Note	2012	2013	2014	2015	Note	2012	2013	2014	2015
05	Mining of coal and lignite		...	...	...	...		...	...	...	...
051	Mining of hard coal		...	...	...	...		...	...	...	...
052	Mining of lignite		...	...	...	...		...	...	...	...
06	Extraction of crude petroleum & natural gas		...	...	...	...		...	...	...	...
061	Extraction of crude petroleum		...	...	...	...		...	...	...	...
062	Extraction of natural gas		...	...	...	...		...	...	...	...
07	Mining of metal ores		...	...	...	...		...	...	...	...
071	Mining of iron ores		...	...	...	...		...	...	...	...
072	Mining of non-ferrous metal ores		...	...	...	...		...	...	...	...
08	Other mining and quarrying		...	68912.5	9861.1	...		...	15412.8	4447.5	...
081	Quarrying of stone, sand and clay		...	67403.6	132.4	...		...	14811.2	108.4	...
089	Mining and quarrying n.e.c.		...	1508.9	9728.7	...		...	601.6	4339.1	...
09	Mining support service activities		...	...	...	...		...	...	...	...
091	Support activities for petroleum		...	...	...	...		...	...	...	...
099	Support activities for other mining, quarrying		...	...	...	...		...	...	...	...
B	Mining and quarrying	a/	...	68912.5	9861.1	...	a/	...	15412.8	4447.5	...
35	Electricity, gas, steam & air conditioning		...	420215.9	437837.8	...		...	241222.7	205754.7	...
351	Electric power generation, transmission		...	419890.5	429046.0	...		...	241158.8	201120.7	...
352	Manufacture of gas		...	...	...	...		...	...	...	...
353	Steam and air conditioning supply		...	325.4	8791.9	...		...	63.8	4634.0	...
D	Electricity, gas, steam & air conditioning	a/	...	420215.9	437837.8	...	a/	...	241222.7	205754.7	...
36	Water collection, treatment and supply		...	...	40928.5	...		...	...	29280.7	...
37	Sewerage		...	...	...	...		...	...	...	...
38	Waste collection,treatment,disposal activities		...	345.7	1479.0	...		...	169.7	426.8	...
381	Waste collection		...	...	...	...		...	...	...	...
382	Waste treatment and disposal		...	345.7	...	...		...	169.7	...	...
383	Materials recovery		...	...	1479.0	...		...	...	426.8	...
39	Remediation activities		...	3919.3	...	...		...	163.0	...	...
E	Water supply;sewerage,waste management		...	...	...	...		...	...	...	...

a/ Sum of available data.

State of Palestine

ISIC Revision 4		Number of enterprises					Number of persons engaged					Wages and salaries paid to employees				
		Note	(number)				Note	(number)				Note	(thousands of US Dollars)			
ISIC	Industry		2011	2012	2013	2014		2011	2012	2013	2014		2011	2012	2013	2014
05	Mining of coal and lignite		...	...	...	...		...	...	...	...		...	...	...	...
06	Extraction of crude petroleum & natural gas		...	...	...	...		...	...	...	...		...	...	...	...
07	Mining of metal ores		...	...	...	...		...	...	...	...		...	...	...	...
08	Other mining and quarrying		247	208	211	222		1766	1476	1721	1247		9045	9702	10362	10182
09	Mining support service activities		...	...	...	...		...	...	...	...		...	...	...	...
B	Mining and quarrying		247	208	211	222		1766	1476	1721	1247		9045	9702	10362	10182
35	Electricity, gas, steam & air conditioning		8	13	13	183		3352	2522	3301	3833		29225	37863	54219	82924
D	Electricity, gas, steam & air conditioning		8	13	13	183		3352	2522	3301	3833		29225	37863	54219	82924
36	Water collection, treatment and supply		419	309a/	313	160a/		925	912a/	1000	1031a/		4598	7824a/	8395	11270a/
37	Sewerage		11b/	...a/	...	...a/		52b/	...a/	...	...a/		325b/	...a/	...	...a/
38	Waste collection,treatment,disposal activities		...b/	23c/	24c/	26c/		...b/	143c/	103c/	73c/		...b/	468c/	327c/	500c/
39	Remediation activities		...	...c/	...c/	...c/		...	...c/	...c/	...c/		...	...c/	...c/	...c/
E	Water supply;sewerage,waste management		...	332	...	186		...	1055	...	1104		...	8292	...	11770

a/ 36 includes 37.
b/ 37 includes 38.
c/ 38 includes 39.

State of Palestine

ISIC Revision 4		Output (valuation not defined)					Value added (valuation not defined)				
		Note	(thousands of US Dollars)				Note	(thousands of US Dollars)			
ISIC	Industry		2011	2012	2013	2014		2011	2012	2013	2014
05	Mining of coal and lignite		...	...	...	...		...	...	...	...
06	Extraction of crude petroleum & natural gas		...	...	...	...		...	...	...	...
07	Mining of metal ores		...	...	...	...		...	...	...	...
08	Other mining and quarrying		101725	73268	120439	74336		68189	33932	66055	39404
09	Mining support service activities		...	...	...	...		...	...	...	...
B	Mining and quarrying		101725	73268	120439	74336		68189	33932	66055	39404
35	Electricity, gas, steam & air conditioning		409567	319202	600648	757851		220758	177683	194852	213426
D	Electricity, gas, steam & air conditioning		409567	319202	600648	757851		220758	177683	194852	213426
36	Water collection, treatment and supply		11296	36033a/	35262	67764a/		1816	14743a/	10058	33421a/
37	Sewerage		5590b/	...a/	...	...a/		616b/	...a/	...	...a/
38	Waste collection,treatment,disposal activities		...b/	3627c/	6588c/	2508c/		...b/	1081c/	2551c/	1323c/
39	Remediation activities		...	...c/	...c/	...c/		...	...c/	...c/	...c/
E	Water supply;sewerage,waste management		...	39660	...	70272		...	15824	...	34743

a/ 36 includes 37.
b/ 37 includes 38.
c/ 38 includes 39.

Swaziland

ISIC Revision 4		Number of establishments					Number of employees					Wages and salaries paid to employees				
			(number)					(number)					(millions of Swazi Emalangeni)			
ISIC	Industry	Note	2011	2012	2013	2014	Note	2011	2012	2013	2014	Note	2011	2012	2013	2014
051	Mining of hard coal		2a/	...	...	...		319	...	...	...		21a/	...	...	...
052	Mining of lignite		...a/	...	...	...		...	...	...	...		...a/	...	...	...
061	Extraction of crude petroleum		...	...	...	...		...	...	...	...		...	...	...	...
062	Extraction of natural gas		...	...	...	...		...	...	...	...		...	...	...	...
071	Mining of iron ores		...	...	...	...		...	...	...	...		...	...	...	...
072	Mining of non-ferrous metal ores		...	...	...	...		...	...	...	...		...	...	...	...
081	Quarrying of stone, sand and clay		6b/	...	...	...		49	...	...	...		...	...	...	...
089	Mining and quarrying n.e.c.		...b/	...	...	...		...	...	...	...		...	...	...	...
091	Support activities for petroleum		...	...	...	...		...	...	...	...		...	...	...	...
099	Support activities for other mining, quarrying		...	...	...	...		...	...	...	...		...	...	...	...
B	Mining and quarrying		8	...	...	...		368	...	...	...		21	...	...	...
351	Electric power generation, transmission		10c/	...	...	...		645	...	...	...		134c/	...	...	...
352	Manufacture of gas		...c/	...	...	...		11	...	...	...		...c/	...	...	...
353	Steam and air conditioning supply		...c/	...	...	...		...	...	...	...		...c/	...	...	...
D	Electricity, gas, steam & air conditioning		10	...	...	...		656	...	...	...		134	...	...	...
360	Water collection, treatment and supply		4	...	...	...		543	...	...	...		79	...	...	...
370	Sewerage		...	...	...	...		...	...	...	...		...	...	...	...
381	Waste collection		11d/	...	...	...		...	...	...	...		8d/	...	...	...
382	Waste treatment and disposal		...d/	...	...	...		54	...	...	...		...d/	...	...	...
383	Materials recovery		...d/	...	...	...		66	...	...	...		...d/	...	...	...
390	Remediation activities		1	...	...	...		...	...	...	...		...	...	...	...
E	Water supply;sewerage,waste management		16	...	...	...		663	...	...	...		87	...	...	...

a/ 051 includes 052.
b/ 081 includes 089.
c/ 351 includes 352 and 353.
d/ 381 includes 382 and 383.

Swaziland

ISIC Revision 4		Output at basic prices					Value added at basic prices				
		Note	(millions of Swazi Emalangeni)				Note	(millions of Swazi Emalangeni)			
ISIC	Industry		2011	2012a/	2013a/	2014a/		2011	2012a/	2013a/	2014a/
051	Mining of hard coal	b/	181	260	371	260	b/	82	118	169	118
052	Mining of lignite	b/	...	...	...	...	b/	...	...	...	...
061	Extraction of crude petroleum		...	...	...	...		...	...	...	...
062	Extraction of natural gas		...	...	...	...		...	...	...	...
071	Mining of iron ores		...	...	...	...		...	...	...	...
072	Mining of non-ferrous metal ores		...	...	...	...		...	...	...	...
081	Quarrying of stone, sand and clay		...	...	...	...		...	...	...	...
089	Mining and quarrying n.e.c.		...	...	...	...		...	...	...	...
091	Support activities for petroleum		...	...	...	...		...	...	...	...
099	Support activities for other mining, quarrying		...	...	...	...		...	...	...	...
B	Mining and quarrying		181	260	371	260		82	118	169	118
351	Electric power generation, transmission	c/	1043	1117	1250	1518	c/	500	535	599	728
352	Manufacture of gas	c/	...	...	...	...	c/	...	...	...	...
353	Steam and air conditioning supply	c/	...	...	...	...	c/	...	...	...	...
D	Electricity, gas, steam & air conditioning		1043	1117	1250	1518		500	535	599	728
360	Water collection, treatment and supply		233	...	...	...		161	...	...	...
370	Sewerage		...	...	...	...		...	...	...	...
381	Waste collection		56d/	...	...	...		43d/	...	...	...
382	Waste treatment and disposal		...d/	...	...	...		...d/	...	...	...
383	Materials recovery		...d/	...	...	...		...d/	...	...	...
390	Remediation activities		...	...	...	...		...	...	...	...
E	Water supply;sewerage,waste management		289	317	374	408		204	224	264	288

a/ Provisional data.
b/ 051 includes 052.
c/ 351 includes 352 and 353.
d/ 381 includes 382 and 383.

Sweden

ISIC Revision 4		Number of enterprises					Number of employees					Wages and salaries paid to employees				
			(number)					(number)					(millions of Swedish Kronor)			
ISIC	Industry	Note	2012	2013	2014	2015	Note	2012	2013	2014	2015	Note	2012	2013	2014	2015
051	Mining of hard coal		...	...	-	-		...	...	-	-		...	...	-	-
052	Mining of lignite		...	...	-	-		...	...	-	-		...	...	-	-
061	Extraction of crude petroleum		...	...	-	-		...	...	-	-		...	...	-	-
062	Extraction of natural gas		...	...	-	-		...	...	-	-		...	...	-	-
071	Mining of iron ores		11	13	15	11		3641	3975	4029	3698		1760	1899	2022	1809
072	Mining of non-ferrous metal ores		24	26	...	27		...	...	...	...		...	...	...	...
081	Quarrying of stone, sand and clay		495	479	470	461		2567	2484	2382	1856		866	808	867	732
089	Mining and quarrying n.e.c.		102	97	93	92		...	239	241	158		...	79	75	51
091	Support activities for petroleum		50	58	59	63		...	49	69	55		...	26	30	31
099	Support activities for other mining, quarrying		71	76	...	78		...	...	...	...		...	...	...	...
B	Mining and quarrying		753	749	740	732		9200	9478	9477	9510		3923	4065	4271	4396
351	Electric power generation, transmission		1867	1957	2128	...		19712	19772	20061	20135		9110	9398	9616	9910
352	Manufacture of gas		67	76	86	82		272	285	281	277		144	153	156	160
353	Steam and air conditioning supply		248	254	256	252		8287	8331	8173	7763		3564	3675	3659	3595
D	Electricity, gas, steam & air conditioning		2182	2287	2470	...		28271	28388	28515	28175		12818	13226	13432	13665
360	Water collection, treatment and supply		218	226	256	259		1302	1448	1626	1601		496	581	661	673
370	Sewerage		200	212	225	232		1905	2105	2174	2321		672	754	807	887
381	Waste collection		280	288	290	285		7941	7921	8484	9017		2598	2694	2989	3294
382	Waste treatment and disposal		134	137	141	144		1516	1425	1431	1468		606	607	613	653
383	Materials recovery		412	426	431	434		4349	4450	4436	4344		1554	1607	1637	1642
390	Remediation activities		117	116	114	108		324	352	345	374		84	97	113	116
E	Water supply;sewerage,waste management		1361	1405	1457	1462		17339	17701	18496	19125		6011	6340	6819	7265

Sweden

ISIC Revision 4		Output (valuation not defined)					Value added at factor values				
			(millions of Swedish Kronor)					(millions of Swedish Kronor)			
ISIC	Industry	Note	2012	2013	2014	2015	Note	2012	2013	2014	2015
051	Mining of hard coal		...	...	-	-		...	...	-	-
052	Mining of lignite		...	...	-	-		...	...	-	-
061	Extraction of crude petroleum		...	...	-	-		...	...	-	-
062	Extraction of natural gas		...	...	-	-		...	...	-	-
071	Mining of iron ores		25440	23605	20030	15030		14052	10653	8140	6257
072	Mining of non-ferrous metal ores		...	...	...	...		...	...	...	...
081	Quarrying of stone, sand and clay		8222	7435	7835	6643		2327	2158	2296	2153
089	Mining and quarrying n.e.c.		...	493	538	413		...	189	208	152
091	Support activities for petroleum		...	1032	872	729		...	606	332	520
099	Support activities for other mining, quarrying		...	...	...	...		...	...	...	...
B	Mining and quarrying		45620	41027	38409	35439		23380	18115	16194	15852
351	Electric power generation, transmission		133880	128449	127472	124160		64301	57912	59009	52029
352	Manufacture of gas		2079	2039	2091	1901		908	805	893	797
353	Steam and air conditioning supply		38916	38068	37054	36099		15823	16373	16747	16943
D	Electricity, gas, steam & air conditioning		174875	168556	166616	162160		81032	75092	76648	69769
360	Water collection, treatment and supply		4960	5202	5714	6097		1816	1886	2131	2114
370	Sewerage		3968	4404	4617	4975		1754	1972	2064	2207
381	Waste collection		13589	13903	16244	18685		5496	6079	6424	7019
382	Waste treatment and disposal		4621	4109	3777	4265		1769	2032	1307	1707
383	Materials recovery		21166	19958	19682	19147		3705	10082	3879	3755
390	Remediation activities		412	457	527	570		152	272	223	247
E	Water supply;sewerage,waste management		48714	48034	50561	53739		14691	22323	16026	17050

Switzerland

ISIC Revision 4		Number of enterprises					Number of persons engaged					Wages and salaries paid to employees				
			(number)					(number)					(millions of Swiss Francs)			
ISIC	Industry	Note	2012	2013	2014	2015	Note	2012	2013	2014	2015	Note	2012	2013	2014	2015
051	Mining of hard coal		...	...	...	-a/		...	...	...	-a/		...	...	...	-a/
052	Mining of lignite		...	...	...	...a/		...	...	...	...a/		...	...	...	...a/
061	Extraction of crude petroleum		1	...	...	-		...	...	...	...		...	...	...	...
062	Extraction of natural gas		-	...	...	1		...	...	...	...		...	...	...	...
071	Mining of iron ores		...	...	...	-b/		...	...	...	-b/		...	...	...	-b/
072	Mining of non-ferrous metal ores		...	...	...	...b/		...	...	...	...b/		...	...	...	...b/
081	Quarrying of stone, sand and clay		180	202	197	197		3696	3946	3947	4058		317	...	349c/	...
089	Mining and quarrying n.e.c.		7	10	8	9		368	349	238	263		32	...	...c/	...
091	Support activities for petroleum		5	4	5	4		71	61	99	100		12	...	15	...
099	Support activities for other mining, quarrying		-	-	1	1		-	-	...	...		-	...	...	...
B	Mining and quarrying		193	216	211	212		4134	4355	4284	4421		361	368	364	383
351	Electric power generation, transmission		311	406	419	403		27048	27724	28004	27436		2653	2720	2770	2715
352	Manufacture of gas		25	33	37	34		937	970	1055	1030		...	...	...	...
353	Steam and air conditioning supply		10	20	20	21		498	46	67	608		...	...	...	...
D	Electricity, gas, steam & air conditioning		346	459	476	458		28484	28740	29126	29074		2798	2827	2885	2875
360	Water collection, treatment and supply		49	95	102	101		1055	1226	1364	1481		89	89	100	108
370	Sewerage		168	170	174	170		2769	2645	2808	2859		226	224	235	240
381	Waste collection		133	129	135	134		2852	2216	2514	3023		202	150	167	212
382	Waste treatment and disposal		98	111	126	129		1570	2424	2680	2525		130	190	213	211
383	Materials recovery		186	189	199	195		4750	4799	4631	3539		360	360	355	272
390	Remediation activities		19	18	20	13		232	231	192	103		13	14	12	7
E	Water supply;sewerage,waste management		653	712	756	742		13228	13540	14188	13529		1020	1026	1082	1051

a/ 051 includes 052.
b/ 071 includes 072.
c/ 081 includes 089.

Switzerland

ISIC Revision 4		Output (valuation not defined)					Value added at factor values				
		Note	(millions of Swiss Francs)				Note	(millions of Swiss Francs)			
ISIC	Industry		2012	2013	2014	2015		2012	2013	2014	2015
051	Mining of hard coal		...	...	...	-a/		...	...	...	-a/
052	Mining of lignite		...	...	...	...a/		...	...	...	...a/
061	Extraction of crude petroleum		...	...	...	...		...	...	...	...
062	Extraction of natural gas		...	...	...	...		...	...	...	...
071	Mining of iron ores		...	...	...	-b/		...	...	...	-b/
072	Mining of non-ferrous metal ores		...	...	...	...b/		...	...	...	...b/
081	Quarrying of stone, sand and clay		1984	...	2144c/	...		758	...	789c/	...
089	Mining and quarrying n.e.c.		169	...	...c/	...		81	...	...c/	...
091	Support activities for petroleum		53	...	39	...		18	...	22	...
099	Support activities for other mining, quarrying		-	...	...	...		-	...	...	...
B	Mining and quarrying		2206	2216	2184	2220		857	861	811	851
351	Electric power generation, transmission		40806	39953	38650	33102		9201	8872	7810	8374
352	Manufacture of gas		...	...	...	...		...	...	...	...
353	Steam and air conditioning supply		...	...	...	...		...	...	...	...
D	Electricity, gas, steam & air conditioning		44285	52664	49034	40298		9622	9364	8295	8925
360	Water collection, treatment and supply		698	646	707	681		268	249	274	279
370	Sewerage		828	813	834	798		435	422	466	469
381	Waste collection		735	775	798	945		392	291	330	402
382	Waste treatment and disposal		841	1392	1424	1352		371	509	612	553
383	Materials recovery		2664	2411	2324	1824		659	623	633	481
390	Remediation activities		31	32	44	24		18	14	18	11
E	Water supply;sewerage,waste management		5798	6070	6132	5624		2143	2108	2334	2196

a/ 051 includes 052.
b/ 071 includes 072.
c/ 081 includes 089.

The former Yugoslav Republic of Macedonia

ISIC Revision 4		Number of establishments					Number of employees					Wages and salaries paid to employees			
		Note	(number)				Note	(number)				Note	(millions of Macedonia Denars)		
ISIC	Industry		2011	2012	2013	2014		2011	2012	2013	2014		2011	2012	2013	2014
051	Mining of hard coal		3	...	...	...		30	...	...	...		20	43a/	30a/	30a/
052	Mining of lignite		2	...	...	...		48	...	...	...		20	...a/	...a/	...a/
061	Extraction of crude petroleum		...	...	...	...		...	...	...	...		...	...	...	...
062	Extraction of natural gas		3	...	...	...		17	...	...	...		8	...	...	...
071	Mining of iron ores		2	...	...	...		223	...	...	...		71	1079b/	1220b/	1268b/
072	Mining of non-ferrous metal ores		9	...	...	...		2308	...	...	...		1017	...b/	...b/	...b/
081	Quarrying of stone, sand and clay		89	...	...	...		1227	...	...	...		469	576c/	509c/	525c/
089	Mining and quarrying n.e.c.		27	...	...	...		402	...	...	...		140	...c/	...c/	...c/
091	Support activities for petroleum		1	...	...	...		21	...	...	...		5	...	...	...
099	Support activities for other mining, quarrying		6	...	...	...		111	...	...	...		38	...	...	...
B	Mining and quarrying		142d/	...	...	...		4387d/	4278	4208	4178		1788d/	1755	1797	1906
351	Electric power generation, transmission		48	...	...	...		7175	...	...	...		5134	5515e/	5601e/	5527e/
352	Manufacture of gas		2	...	...	...		11	...	...	...		5	...e/	...e/	...e/
353	Steam and air conditioning supply		12	...	...	...		684	...	...	...		304	...e/	...e/	...e/
D	Electricity, gas, steam & air conditioning		62d/	...	...	...		7870d/	7951	8183	7974		5443d/	5515	5601	5527
360	Water collection, treatment and supply		78	...	...	...		5895	...	...	...		1997	2091	2086	2106
370	Sewerage		4	...	...	...		14	...	...	...		5	...	...	...
381	Waste collection		72	...	...	...		2231	...	...	...		709	1011f/	1098f/	1218f/
382	Waste treatment and disposal		27	...	...	...		517	...	...	...		177	...f/	...f/	...f/
383	Materials recovery		80	...	...	...		409	...	...	...		108	...f/	...f/	...f/
390	Remediation activities		7	...	...	...		44	...	...	...		11	...	...	...
E	Water supply;sewerage,waste management		268d/	...	...	...		9110d/	9385	9661	10193		3007d/	3131	3213	3356

a/ 051 includes 052.
b/ 071 includes 072.
c/ 081 includes 089.
d/ Sum of available data.
e/ 351 includes 352 and 353.
f/ 381 includes 382 and 383.

The former Yugoslav Republic of Macedonia

ISIC Revision 4			Output at basic prices					Value added at basic prices			
			(millions of Macedonia Denars)					(millions of Macedonia Denars)			
ISIC	Industry	Note	2011	2012	2013	2014	Note	2011	2012	2013	2014
051	Mining of hard coal		367	232a/	174a/	171a/		139	106a/	65a/	60a/
052	Mining of lignite		77	...a/	...a/	...a/		33	...a/	...a/	...a/
061	Extraction of crude petroleum		...	...	...	...		...	...	...	...
062	Extraction of natural gas		42	...	...	...		22	...	...	...
071	Mining of iron ores		523	8606b/	8892b/	8761b/		313	4879b/	5210b/	3921b/
072	Mining of non-ferrous metal ores		7607	...b/	...b/	...b/		4488	...b/	...b/	...b/
081	Quarrying of stone, sand and clay		2214	2683c/	3215c/	3376c/		1182	1332c/	1450c/	1519c/
089	Mining and quarrying n.e.c.		959	...c/	...c/	...c/		535	...c/	...c/	...c/
091	Support activities for petroleum		64	...	...	...		2	...	...	...
099	Support activities for other mining, quarrying		375	...	...	...		200	...	...	...
B	Mining and quarrying		12228	12116	12672	12788		6914	6610	6876	6632
351	Electric power generation, transmission		26854	33826d/	35647d/	35810d/		13004	12537d/	14303d/	14262d/
352	Manufacture of gas		37	...d/	...d/	...d/		21	...d/	...d/	...d/
353	Steam and air conditioning supply		4767	...d/	...d/	...d/		719	...d/	...d/	...d/
D	Electricity, gas, steam & air conditioning		31658	33826	35647	35810		13744	12537	14303	14262
360	Water collection, treatment and supply		3976	4438	4024	4551		2666	3037	2646	3157
370	Sewerage		13	...	...	...		9	...	...	...
381	Waste collection		1875	3420e/	3343e/	3639e/		1187	1842e/	1746e/	1961e/
382	Waste treatment and disposal		558	...e/	...e/	...e/		198	...e/	...e/	...e/
383	Materials recovery		894	...e/	...e/	...e/		369	...e/	...e/	...e/
390	Remediation activities		63	...	...	...		42	...	...	...
E	Water supply;sewerage,waste management		7379	7932	7481	8309		4471	4937	4460	5188

a/ 051 includes 052.
b/ 071 includes 072.
c/ 081 includes 089.
d/ 351 includes 352 and 353.
e/ 381 includes 382 and 383.

Turkey

ISIC Revision 4		Number of enterprises					Number of employees					Wages and salaries paid to employees			
			(number)					(number)					(millions of Turkish Liras)		
ISIC	Industry	Note	2012	2013	2014	2015	Note	2012	2013	2014	2015	Note	2012	2013	2014	2015
051	Mining of hard coal		71	78	87	83		18224	16532	15607	14704		566.1	589.0	609.0	687.6
052	Mining of lignite		242	283	240	224		32527	32016	28731	25079		779.5	750.2	697.2	851.1
061	Extraction of crude petroleum		8	10	10	11	a/	5446	6362	6723	6670	a/	416.8	515.3	568.9	544.9
062	Extraction of natural gas		6	8	8	9	a/	...	...	...	...	a/	...	...	...	...
071	Mining of iron ores		34	45	48	45		2322	2752	3002	4253		68.2	106.7	145.0	178.6
072	Mining of non-ferrous metal ores		339	393	405	361		19335	20067	20803	18908		496.2	576.4	741.1	754.6
081	Quarrying of stone, sand and clay		2154	2366	2188	2044		34523	42726	39934	41992		489.2	660.9	780.7	922.8
089	Mining and quarrying n.e.c.		472	626	638	683		13199	14777	13982	14432		321.9	390.6	408.0	484.6
091	Support activities for petroleum		28	40	47	50		2014	1170	1334	1060		97.3b/	70.3	86.1	82.5
099	Support activities for other mining, quarrying		12	29	44	48		161	380	438	548		...b/	12.5	9.8	12.2
B	Mining and quarrying		3366	3878	3715	3558		127751	136782	130554	127646		3235.1	3671.8	4045.8	4518.9
351	Electric power generation, transmission		546	1012	1322	1495		63443	72291	77796	84085		2655.5	3421.9	3823.9	4156.3
352	Manufacture of gas		71	88	103	112		6045	6475	6933	7829		284.8	344.8	401.1	493.2
353	Steam and air conditioning supply		54	71	88	96		604	710	834	869		12.8	16.6	19.4	23.3
D	Electricity, gas, steam & air conditioning		671	1171	1513	1703		70092	79476	85563	92783		2953.1	3783.3	4244.4	4672.8
360	Water collection, treatment and supply		2817	2242	1386	1063		33204	34048	36660	36647		1363.2	1498.0	1839.2	2037.6
370	Sewerage		154	216	276	288		1324	2959	2871	3437		32.8	54.9	69.4	91.1
381	Waste collection		279	353	551	653		50137	50154	51182	60951		848.3	967.1	1124.9	1550.6
382	Waste treatment and disposal		35	59	103	117		1287	1599	2204	2972		35.6	46.7	64.2	85.1
383	Materials recovery		407	651	1000	1147		7453	8500	9032	8244		110.3	138.4	156.0	159.0
390	Remediation activities		14	14	17	16		173	776	563	311		3.5	13.1	9.5	7.0
E	Water supply;sewerage,waste management		3706	3535	3333	3284		93578	98036	102512	112562		2393.9	2718.1	3263.2	3930.4

a/ 061 includes 062.
b/ 091 includes 099.

Turkey

ISIC Revision 4		Output (valuation not defined)					Value added at factor values				
			(millions of Turkish Liras)					(millions of Turkish Liras)			
ISIC	Industry	Note	2012	2013	2014	2015	Note	2012	2013	2014	2015
051	Mining of hard coal		694.0	619.7	641.4	535.6		281.5	276.6	241.5	165.4
052	Mining of lignite		4853.9	3922.6	3693.3	4246.8		2108.5	1696.6	843.7	1484.2
061	Extraction of crude petroleum		3343.5a/	4559.5a/	3532.4	2438.5	a/	1711.6	2386.9	2710.7	526.4
062	Extraction of natural gas		...a/	...a/	79.9	80.8	a/	...	...	...	...
071	Mining of iron ores		1022.8	1116.0	1375.9	1362.4		486.7	403.9	459.1	517.5
072	Mining of non-ferrous metal ores		6281.5	7566.8	7337.7	6903.1		3616.2	3805.0	4029.0	3304.0
081	Quarrying of stone, sand and clay		5308.4	6936.4	7490.8	8474.9		1204.2	1789.0	1825.5	2352.8
089	Mining and quarrying n.e.c.		3395.4	4268.2	4603.1	5220.3		1638.2	1913.0	2171.0	2525.5
091	Support activities for petroleum		595.0	453.6	769.6	555.7		266.7	108.7b/	248.1	100.9
099	Support activities for other mining, quarrying		12.5	64.6	63.5	50.8		0.9	...b/	17.1	17.2
B	Mining and quarrying		25507.1	29507.5	29587.6	29869.0		11314.5	12379.7	12545.9	10993.9
351	Electric power generation, transmission		109237.7	136430.0	155911.0	168366.0		14062.1	15801.4	15147.6	23917.8
352	Manufacture of gas		11896.7	18997.5	20592.1	27393.5		950.0	1240.9	1234.5	2067.3
353	Steam and air conditioning supply		92.7	97.4	115.7	148.8		42.7	34.4	45.1	56.5
D	Electricity, gas, steam & air conditioning		121227.1	155525.0	176618.8	195909.0		15054.8	17076.7	16427.1	26041.7
360	Water collection, treatment and supply		9606.7	11190.6	11995.4	14734.1		5488.0	4809.6	6092.2	7053.6
370	Sewerage		323.8	480.8	509.2	414.3		...	95.4	115.0	118.6
381	Waste collection		2899.6	3129.5	3801.9	4711.9		1383.8	1421.4	1637.6	2211.3
382	Waste treatment and disposal		236.4	664.3	1657.9	707.1		91.1	168.0	182.2	151.7
383	Materials recovery		3948.0	3820.3	3631.6	3118.1		240.6	330.5	325.9	312.5
390	Remediation activities		17.7	46.3	36.8	29.6		...	26.7	21.9	10.5
E	Water supply;sewerage,waste management		17032.1	19331.8	21632.9	23715.2		7233.1	6851.7	8374.7	9858.3

a/ 061 includes 062.
b/ 091 includes 099.

Ukraine

ISIC Revision 4		Number of establishments					Number of employees					Wages and salaries paid to employees				
			(number)					(number)					(millions of Ukrainian Hryvnias)			
ISIC	Industry	Note	2012	2013	2014	2015	Note	2012	2013	2014	2015	Note	2012	2013	2014	2015
051	Mining of hard coal		365	...a/	101	151		...a/	...a/	...a/	138765		...a/	...a/	...a/	9909
052	Mining of lignite		2	...a/	2	6		...a/	...a/	...a/	63		...a/	...a/	...a/	1
061	Extraction of crude petroleum		39	41	40	33		28253	27246	27022	26721		1376	1329	1591	1902
062	Extraction of natural gas		62	78	70	82		27445	28371	23169	23200		1984	2053	1943	1552
071	Mining of iron ores		29	31	21	21		53079	52916	52512	51244		3008	3163	3689	4292
072	Mining of non-ferrous metal ores		19	28	25	23		19485	18536	17839	23531		863	884	963	1565
081	Quarrying of stone, sand and clay		918	1097	928	933		38716	36358	31832	29850		1311	1283	1196	1317
089	Mining and quarrying n.e.c.		80	112	109	118		5719	5179	2056	4757		275	288	71	281
091	Support activities for petroleum		56	82	76	80		1445	1911	1744	2364		101	121	171	249
099	Support activities for other mining, quarrying		52	50	36	38		2578	2639	1198	1227		112	109	70	92
B	Mining and quarrying		1622	1918	1408	1485		482587	473841	315713	301722		27827	29207	20908	21161
351	Electric power generation, transmission		538	819	682	678		224247	222180	211785	191700		13035	14374	15228	15359
352	Manufacture of gas		216	235	203	233		88879	87718	76044	68367		3015	3055	2764	2916
353	Steam and air conditioning supply		580	669	605	711		108166	106823	74782	80020		4288	4595	3053	4136
D	Electricity, gas, steam & air conditioning		1334	1723	1490	1622		421292	416721	362611	340087		20338	22024	21045	22410
360	Water collection, treatment and supply		1651	1786	1561	1577		102914	100137	83558	82233		3048	3182	2931	3426
370	Sewerage		211	257	208	230		14129	19738	17428	19451		403	621	578	754
381	Waste collection		912	1093	882	889		28038	25956	18841	18180		786	753	604	651
382	Waste treatment and disposal		102	132	130	154		1873	1763	2240	2163		52	60	81	85
383	Materials recovery		735	784	592	560		12506	10711	7219	6273		308	350	238	244
390	Remediation activities		32	38	38	47		706	383	724	702		20	10	39	39
E	Water supply;sewerage,waste management		3643	4090	3411	3457		160166	158688	130010	129002		4617	4976	4470	5199

a/ Data suppressed due to confidentiality rules.

Ukraine

ISIC Revision 4		Output at factor values					Value added at producers' prices				
			(millions of Ukrainian Hryvnias)					(millions of Ukrainian Hryvnias)			
ISIC	Industry	Note	2012	2013	2014	2015	Note	2012	2013	2014	2015
051	Mining of hard coal	a/	...	...	...	38788		...	...	...	...
052	Mining of lignite	a/	...	...	...	11		...	...	...	...
061	Extraction of crude petroleum		14776	14578	25365	24900		11317	6410	15989	...
062	Extraction of natural gas		20080	21547	24659	48716		11174	12255	14226	...
071	Mining of iron ores		51281	58083	78905	85407		25683	31720	46432	...
072	Mining of non-ferrous metal ores		4316	3616	4028	9098		2576	1931	1804	...
081	Quarrying of stone, sand and clay		11290	10570	13787	16724		4610	4414	7203	...
089	Mining and quarrying n.e.c.		1561	1447	684	1890		928	848	429	...
091	Support activities for petroleum		810	834	1428	2129		356	97	672	...
099	Support activities for other mining, quarrying		724	790	823	1283		87	118	163	...
B	Mining and quarrying		157921	170060	184404	228945		94292	93376	110908	...
351	Electric power generation, transmission		107387	107321	135454	160314		49264	47020	70595	...
352	Manufacture of gas		15962	17635	17162	19241		9452	12106	10434	...
353	Steam and air conditioning supply		27794	24234	17020	31570		7170	408	8680	...
D	Electricity, gas, steam & air conditioning		151143	149190	169635	211125		65887	59534	89709	...
360	Water collection, treatment and supply		9885	9055	8881	10977		7294	4887	4327	...
370	Sewerage		1121	1684	1661	2266		645	893	1012	...
381	Waste collection		2946	3079	2836	3052		1574	1490	1318	...
382	Waste treatment and disposal		223	256	440	365		107	112	299	...
383	Materials recovery		7196	6977	7083	6363		1216	1234	1899	...
390	Remediation activities		106	54	95	208		58	19	65	...
E	Water supply;sewerage,waste management		21477	21106	20996	23231		10894	8636	8921	...

a/ Data suppressed due to confidentiality rules.

United Kingdom

ISIC Revision 4		Number of enterprises					Number of employees					Wages and salaries paid to employees				
		Note	(number)				Note	(number)				Note	(millions of British Pounds)			
ISIC	Industry		2012	2013	2014	2015		2012	2013	2014	2015		2012	2013	2014	2015
051	Mining of hard coal		...	20	...	...		7962	5421	3777	2840		...	214	...	...
052	Mining of lignite		-	-	-	-		-	-	-	-		-	-	-	-
061	Extraction of crude petroleum		98	92	88	140a/		14239	13694	...	16179		1876a/	2070a/	2259	2193a/
062	Extraction of natural gas		53	66	57	...a/		860	666	...	928		...a/	...a/	22	...a/
071	Mining of iron ores		-	7b/	-	-		-	...	-	-		-	-b/	-	-
072	Mining of non-ferrous metal ores		...	...b/	...	...		...	...	...	...		...	...b/	...	...
081	Quarrying of stone, sand and clay		458	434	423	416		22922	17376	16649	20369c/		480	642	582	710
089	Mining and quarrying n.e.c.		245	275	267	283		3861	2136	2598	...c/		81	93	206	127
091	Support activities for petroleum		251	286	284	251		...	...	26863d/	...		1424	1575	1706	1753
099	Support activities for other mining, quarrying		127	123	120	125		...	...	...d/	...		1	10	22	40
B	Mining and quarrying		1263	1303	1263	1234		74827	65431	66567	68544		4164	4604	4956	4897
351	Electric power generation, transmission		1710	2459	3151	4150		80214	86956	88690	94677		3125	3074	3164	3196
352	Manufacture of gas		98	...	112	109		...	...	...	...		1628	...	1581	1602
353	Steam and air conditioning supply		21	...	22	20		...	...	...	...		-	...	2	2
D	Electricity, gas, steam & air conditioning		1829	2577	3285	4279		120087	129454	128770	139738		4754	4708	4746	4800
360	Water collection, treatment and supply		120	106	99	111		...	40143	...	...		1164	1295	1437	1402
370	Sewerage		899	924	921	970		14709	18031	13948	19013		322	347	409	388
381	Waste collection		2358	2377	2365	2433		51387	47907	47277	51783		1126	1197	1260	1452
382	Waste treatment and disposal		946	988	993	1037		24483	25015	32282	21081		643	652	776	569
383	Materials recovery		1884	1982	1903	1859		27335	28490	25796	26768		629	687	612	598
390	Remediation activities		603	696	769	795		...	4030	...	...		57	57	127	92
E	Water supply;sewerage,waste management		6810	7073	7050	7205		159688	163617	162145	166129		3942	4235	4622	4502

a/ 061 includes 062.
b/ 071 includes 072.
c/ 081 includes 089.
d/ 091 includes 099.

United Kingdom

ISIC Revision 4		Output (valuation not defined)					Value added at factor values				
			(millions of British Pounds)					(millions of British Pounds)			
ISIC	Industry	Note	2012	2013	2014	2015	Note	2012	2013	2014	2015
051	Mining of hard coal		...	...	...	...		...	...	...	...
052	Mining of lignite		-	-	-	-		-	-	-	-
061	Extraction of crude petroleum		36183a/	36841a/	28981	22737a/		19650a/	18451a/	11413	9727a/
062	Extraction of natural gas		...a/	...a/	192	...a/		...a/	...a/	131	...a/
071	Mining of iron ores		-	...	-	-		-	...	-	-
072	Mining of non-ferrous metal ores		...	...	...	...		...	...	...	...
081	Quarrying of stone, sand and clay		3502	5274	5071	5268		937	1556	1596	1692
089	Mining and quarrying n.e.c.		475	550	489	641		237	228	190	203
091	Support activities for petroleum		6380	7532	7652	7128		3052	3134	2473	2855
099	Support activities for other mining, quarrying		10	13	25	114		8	-18	-1	39
B	Mining and quarrying		47801	51060	43127	36376		24253	23655	16125	14747
351	Electric power generation, transmission		37643	38197	40036	39590		17002	16267	16600	16515
352	Manufacture of gas		10422	...	...	...		6173	...	...	...
353	Steam and air conditioning supply		2	...	...	...		1	...	...	...
D	Electricity, gas, steam & air conditioning		48067	49762	52760	51165		23177	22342	23353	22268
360	Water collection, treatment and supply		10914	12092	12382	12244		7939	8997	9318	8961
370	Sewerage		2778	2816	3115	4488		1828	1570	1832	2915
381	Waste collection		6440	6741	7945	8506		2675	2571	2913	3315
382	Waste treatment and disposal		3165	3933	4061	3833		1136	1326	1646	1655
383	Materials recovery		7931	6965	7311	5153		1958	1398	1741	1383
390	Remediation activities		304	376	733	627		196	164	438	457
E	Water supply;sewerage,waste management		31531	32923	35547	34852		15733	16026	17889	18685

a/ 061 includes 062.

United States of America

ISIC Revision 4		Number of enterprises					Number of employees					Wages and salaries paid to employees				
			(number)					(number)					(millions of US Dollars)			
ISIC	Industry	Note	2012	2013	2014	2015	Note	2012	2013	2014	2015	Note	2012	2013	2014	2015
051	Mining of hard coal		2169	2155	2172	2085		28172	27445	30186	28248		3578	3403	4003	3165
052	Mining of lignite		513	473	457	421		97057	91642	82776	76441		8255	7791	7323	6631
061	Extraction of crude petroleum		2131	2118	2142	2053		24793	38469	38745	52990		2796	5306	5996	8041
062	Extraction of natural gas		2263	2223	2258	2159		77508	77045	77059	59133		12301	11146	11658	8384
071	Mining of iron ores		16	11	11	12		...	...	...	...		...	...	...	...
072	Mining of non-ferrous metal ores		297	282	277	232		41732	44653	44944	42516		3857	4429	4313	4187
081	Quarrying of stone, sand and clay		2378	2296	2310	2285		55505	58879	62318	61049		3272	3635	4136	4161
089	Mining and quarrying n.e.c.		545	532	537	510		18615	16872	18783	22490		1186	1046	1245	1632
091	Support activities for petroleum		10852	10945	11701	11428		438094	408813	448258	460944		36072	34700	39348	34247
099	Support activities for other mining, quarrying		321	309	305	297		8411	7320	5854	4659		597	485	445	365
B	Mining and quarrying		21485	21344	22170	21482		795928	776603	814667	811795		72529	72479	78994	71123
351	Electric power generation, transmission		1741	1723	1723	1701		537250	532284	531594	526381		51460	52228	53533	55342
352	Manufacture of gas		414	391	395	390		74335	73592	70803	68223		6393	6297	6324	6218
353	Steam and air conditioning supply		342	333	308	300		6708	6405	6179	5920		280	266	270	309
D	Electricity, gas, steam & air conditioning		2497	2447	2426	2391		618293	612281	608576	600524		58133	58790	60126	61869
360	Water collection, treatment and supply		1618	1627	1635	1636		11554	18169	11395	17604		542	...	552	...
370	Sewerage		4633	4666	4739	4763		35656	37131	39072	40774		1521	1653	1816	1936
381	Waste collection		7681	7716	7576	7503		195479	206623	208813	214154		9857	10569	11260	12038
382	Waste treatment and disposal		1241	1127	1129	1123		53259	52418	54654	55976		3395	3198	3514	3577
383	Materials recovery		1131	1158	1161	1110		17880	20002	20060	19748		658	869	864	852
390	Remediation activities		3192	3237	3315	3439		63551	65373	64476	65840		3490	3642	3658	3851
E	Water supply;sewerage,waste management		19496	19531	19555	19574		377379	399716	398470	414096		19464	21027	21664	23296

United States of America

ISIC Revision 4		Output (valuation not defined)					Value added (valuation not defined)				
		Note	(millions of US Dollars)				Note	(millions of US Dollars)			
ISIC	Industry		2012	2013	2014	2015		2012	2013	2014	2015
05	Mining of coal and lignite		42254	...	...	...		38156	...	...	...
06	Extraction of crude petroleum & natural gas		310960	...	...	...		...	...	...	...
07	Mining of metal ores		29096	...	...	...		24912	...	...	...
08	Other mining and quarrying		26937	...	...	...		22189	...	...	...
09	Mining support service activities		119992	...	...	...		119039	...	...	...
B	Mining and quarrying		529240	...	...	...		...	...	...	...
35	Electricity, gas, steam & air conditioning		521076	...	...	...		...	...	...	...
D	Electricity, gas, steam & air conditioning		521076	...	...	...		...	...	...	...
36	Water collection, treatment and supply		9438	...	...	...		...	...	...	...
37	Sewerage		1377	...	...	...		...	...	...	...
38	Waste collection,treatment,disposal activities		...	...	...	...		...	...	...	...
39	Remediation activities		...	...	...	...		...	...	...	...
E	Water supply;sewerage,waste management		...	...	...	...		...	...	...	...

Uruguay

ISIC Revision 4		Number of establishments					Number of persons engaged					Wages and salaries paid to employees				
			(number)					(number)					(millions of Uruguayan Pesos)			
ISIC	Industry	Note	2011	2012	2013	2014	Note	2011	2012	2013	2014	Note	2011	2012	2013	2014
05	Mining of coal and lignite		...	...	...	...		...	...	...	...		...	...	...	...
06	Extraction of crude petroleum & natural gas		...	...	...	...		...	...	...	...		...	...	...	...
07	Mining of metal ores		...	...	...	...		...	...	...	...		...	...	...	...
08	Other mining and quarrying		...	26a/	...	...		...	1007a/	...	...		...	352.1a/	...	...
09	Mining support service activities		...	...a/	...	...		...	...a/	...	...		...	...a/	...	...
B	Mining and quarrying		...	26	...	...		...	1007	...	...		...	352.1	...	...
35	Electricity, gas, steam & air conditioning		9	9	...	...		6725	6796	...	...		5013.0	6516.3	...	...
D	Electricity, gas, steam & air conditioning		9	9	...	...		6725	6796	...	...		5013.0	6516.3	...	...
36	Water collection, treatment and supply		3	3b/	...	...		4496	5100b/	...	...		2234.0	2474.0b/	...	...
37	Sewerage		91c/	...b/	...	...		3309c/	...b/	...	...		588.7c/	...b/	...	...
38	Waste collection,treatment,disposal activities		...c/	91d/	...	...		...c/	3587d/	...	...		...c/	916.8d/	...	...
39	Remediation activities		...c/	...d/	...	...		...c/	...d/	...	...		...c/	...d/	...	...
E	Water supply;sewerage,waste management		94	94	...	...		7805	8686	...	...		2822.7	3390.8	...	...

a/ 08 includes 09.
b/ 36 includes 37.
c/ 37 includes 38 and 39.
d/ 38 includes 39.

Uruguay

ISIC Revision 4		Output at producers' prices					Value added at producers' prices				
			(millions of Uruguayan Pesos)					(millions of Uruguayan Pesos)			
ISIC	Industry	Note	2011	2012	2013	2014	Note	2011	2012	2013	2014
05	Mining of coal and lignite		...	...	...	...		...	...	...	...
06	Extraction of crude petroleum & natural gas		...	...	...	...		...	...	...	...
07	Mining of metal ores		...	...	...	...		...	...	...	...
08	Other mining and quarrying		...	2059.8a/	...	...		...	967.2a/	...	...
09	Mining support service activities		...	...a/	...	...		...	...a/	...	...
B	Mining and quarrying		...	2059.8	...	...		...	967.2	...	...
35	Electricity, gas, steam & air conditioning		28590.8	34929.7	...	...		13192.3	5770.3	...	...
D	Electricity, gas, steam & air conditioning		28590.8	34929.7	...	...		13192.3	5770.3	...	...
36	Water collection, treatment and supply		7263.8	9811.0b/	...	...		4590.6	4678.7b/	...	...
37	Sewerage		3723.7c/	...b/	...	...		1435.5c/	...b/	...	...
38	Waste collection,treatment,disposal activities		...c/	3247.9d/	...	...		...c/	1431.2d/	...	...
39	Remediation activities		...c/	...d/	...	...		...c/	...d/	...	...
E	Water supply;sewerage,waste management		10987.5	13058.9	...	...	e/	6026.1	6110.0	...	...

a/ 08 includes 09.
b/ 36 includes 37.
c/ 37 includes 38 and 39.
d/ 38 includes 39.
e/ Sum of available data.

Viet Nam

ISIC Revision 4		Number of enterprises					Number of persons engaged					Wages and salaries paid to employees				
			(number)					(number)					(billions of Vietnamese Dongs)			
ISIC	Industry	Note	2012	2013	2014	2015	Note	2012	2013	2014	2015 a/	Note	2012	2013	2014	2015
051	Mining of hard coal		72	70	71	86		104614	104055	100822	96006		10372	9927	11370	11226
052	Mining of lignite		12	12	19	9		289	282	243	88		14	12	15	4
061	Extraction of crude petroleum		5	6	8	8		8536	9538	9602	7673		2121	4229	3383	4128
062	Extraction of natural gas		4	2	2	3		1234	1241	1255	1375		343	403	...	373
071	Mining of iron ores		125	125	117	97		5676	5968	6696	5356		331	314	480	357
072	Mining of non-ferrous metal ores		181	170	152	136		13136	12981	9952	8883		828	773	549	480
081	Quarrying of stone, sand and clay		2000	1952	1921	1903		50823	46610	47651	48351		2466	2488	2994	3497
089	Mining and quarrying n.e.c.		123	108	121	123		7281	7321	7808	7061		769	636	749	759
091	Support activities for petroleum		29	48	49	44		2499	3296	3590	2661		964	1410	2294	1065
099	Support activities for other mining, quarrying		51	64	72	61		6206	5662	6741	1862		...	534	611	158
B	Mining and quarrying		2602	2557	2569	2510		200294	196954	195664	180087		18498	20725	22884	22109
351	Electric power generation, transmission		321	321	365	399		196111	214017	202460	120448		18100	27302	31956	18103
352	Manufacture of gas		15	9	13	18		365	432	362	536		20	33	36	46
353	Steam and air conditioning supply		750	753	779	788		7279	7490	7604	7870		274	309	361	449
D	Electricity, gas, steam & air conditioning		1086	1083	1157	1205		203755	221939	210426	128854		18394	27644	32353	18599
360	Water collection, treatment and supply		364	381	423	447		32586	33950	36691	39314		2712	3233	3296	3940
370	Sewerage		189	170	231	287		7638	6696	8107	7972		653	585	759	977
381	Waste collection		346	354	403	433		42064	45550	47847	44356		2998	3406	3559	3528
382	Waste treatment and disposal		93	90	123	131		3088	4186	3855	4921		209	277	259	362
383	Materials recovery		108	103	127	142		1657	1744	2069	2251		55	82	123	136
390	Remediation activities		33	27	40	57		361	264	408	573		21	15	23	38
E	Water supply;sewerage,waste management		1133	1125	1347	1497		87394	92390	98977	99387		6649	7598	8020	8981

a/ Number of employees.

Viet Nam

ISIC Revision 4		Output (valuation not defined)					Value added at producers' prices				
		Note	(billions of Vietnamese Dongs)				Note	(billions of Vietnamese Dongs)			
ISIC	Industry		2012	2013a/	2014	2015		2012	2013	2014	2015
051	Mining of hard coal		70210b/	78867b/	...	...		24064	24371	38753	25030
052	Mining of lignite		...b/	...b/	...	...		43	35	22	4
061	Extraction of crude petroleum		268390c/	287863c/	...	...		122464	104638	86560	37763
062	Extraction of natural gas		...c/	...c/	...	...		50328	46848	55073	41911
071	Mining of iron ores		10886d/	11488d/	...	...		1811	2451	3317	1189
072	Mining of non-ferrous metal ores		...d/	...d/	...	...		3587	3618	3784	2593
081	Quarrying of stone, sand and clay		28407e/	28740e/	...	...		9748	10553	14110	12743
089	Mining and quarrying n.e.c.		...e/	...e/	...	...		1892	1893	2204	2199
091	Support activities for petroleum		...	...	...	...		16296	17680	17321	33591
099	Support activities for other mining, quarrying		...	...	...	...		732	520	1096	748
B	Mining and quarrying		384851	413785	...	...		230965	212606	222539	157816
351	Electric power generation, transmission		...	...	...	...		186617	220158	361009	250454
352	Manufacture of gas		...	...	...	...		775	610	956	593
353	Steam and air conditioning supply		...	...	...	...		8341	1733	2265	2069
D	Electricity, gas, steam & air conditioning		...	...	...	...		195733	222501	364230	253116
360	Water collection, treatment and supply		11801	13824	...	...		7627	8748	13184	13404
370	Sewerage		2387	3217	...	...		1328	1510	1102	1943
381	Waste collection		9921f/	12942f/	...	...		4417	5683	6831	5595
382	Waste treatment and disposal		...f/	...f/	...	...		849	995	997	1476
383	Materials recovery		...f/	...f/	...	...		419	913	1375	1607
390	Remediation activities		144	181	...	...		52	95	87	97
E	Water supply;sewerage,waste management		24252	30164	...	...		14693	17944	23577	24122

a/ Provisional data.
b/ 051 includes 052.
c/ 061 includes 062.
d/ 071 includes 072.
e/ 081 includes 089.
f/ 381 includes 382 and 383.

Yemen

ISIC Revision 3		Number of establishments					Number of employees					Wages and salaries paid to employees				
			(number)					(number)					(millions of Yemeni Rials)			
ISIC	Industry	Note	2010	2011	2012	2013	Note	2010	2011	2012	2013	Note	2010	2011	2012	2013
101	Mining and agglomeration of hard coal		...	...	...	...		...	...	...	...		...	...	...	...
102	Mining and agglomeration of lignite		...	...	...	...		...	...	...	...		...	...	...	...
103	Extraction and agglomeration of peat		...	...	...	...		...	...	...	...		...	...	...	...
111	Extraction of crude oil and natural gas		...	...	...	...		...	...	...	...		...	...	...	...
112	Oil, gas service activities		...	...	...	...		...	...	...	...		...	...	...	...
120	Mining of uranium and thorium ores		...	...	...	...		...	...	...	...		...	...	...	...
131	Iron ores		...	...	...	...		...	...	...	...		...	...	...	...
132	Non-ferrous metal ores,excl.uranium,thorium		...	...	...	...		...	...	...	...		...	...	...	...
141	Quarrying of stone, sand and clay		...	286	343	...	a/	3535	2923	3416	3822	a/	1314	1086	1269	1420
142	Mining and quarrying n.e.c.		...	6	...	...	a/	...	...	...	...	a/	...	...	...	...
C	Total mining and quarrying		...	292	...	...		...	...	...	...		...	...	...	...
401	Electricity production, collection, distribution		...	659	721	...		...	...	...	...		...	...	...	...
402	Gas production and distribution		...	...	...	...		...	...	...	...		...	...	...	...
403	Steam and hot water supply		...	...	...	...		...	...	...	...		...	...	...	...
410	Water collection, purification, distribution		...	1797	2225	...		...	...	...	...		...	...	...	...
E	Total electricity, gas and water supply		...	2456	2946	...		25290	24342	32112	74212		23849	28511	24970	44352

a/ 141 includes 142.

Yemen

ISIC Revision 3		Output (valuation not defined)					Value added (valuation not defined)				
			(millions of Yemeni Rials)					(millions of Yemeni Rials)			
ISIC	Industry	Note	2010	2011	2012	2013	Note	2010	2011	2012	2013
101	Mining and agglomeration of hard coal		...	...	...	...		...	...	...	...
102	Mining and agglomeration of lignite		...	...	...	...		...	...	...	...
103	Extraction and agglomeration of peat		...	...	...	...		...	...	...	...
111	Extraction of crude oil and natural gas	a/	1849096	2006336	1754021	2019213	a/	1593347	1797203	1523340	1784328
112	Oil, gas service activities	a/	...	...	...	...	a/	...	...	...	...
120	Mining of uranium and thorium ores		...	...	...	...		...	...	...	...
131	Iron ores		...	...	...	...		...	...	...	...
132	Non-ferrous metal ores,excl.uranium,thorium		...	...	...	...		...	...	...	...
141	Quarrying of stone, sand and clay	b/	18915	17278	17826	20201	b/	12072	11027	11377	12893
142	Mining and quarrying n.e.c.	b/	...	...	...	...	b/	...	...	...	...
C	Total mining and quarrying		1868011	2023614	1771847	2039414		1605419	1808230	1534717	1797221
401	Electricity production, collection, distribution		...	...	...	...		...	...	...	...
402	Gas production and distribution		...	...	...	...		...	...	...	...
403	Steam and hot water supply		...	...	...	...		...	...	...	...
410	Water collection, purification, distribution		...	...	...	...		...	...	...	...
E	Total electricity, gas and water supply		159999	149902	191178	210410		53226	54114	61957	57243

a/ 111 includes 112.
b/ 141 includes 142.

Zimbabwe

ISIC Revision 3		Number of establishments					Employees					Wages and salaries				
		Note	(number)				Note	(number)				Note	(thousands of US Dollars)			
ISIC	Industry		2010	2011	2012	2013		2010	2011	2012	2013		2010	2011	2012	2013
101	Mining and agglomeration of hard coal		1	1	1	1		...	...	...	...		...	...	...	...
102	Mining and agglomeration of lignite		...	...	...	...		...	...	...	...		...	...	...	...
103	Extraction and agglomeration of peat		...	...	...	...		...	...	...	...		...	...	...	...
111	Extraction of crude oil and natural gas		...	...	...	...		...	...	...	...		...	...	...	...
112	Oil, gas service activities		...	...	...	...		...	...	...	...		...	...	...	...
120	Mining of uranium and thorium ores		...	...	...	...		...	...	...	...		...	...	...	...
131	Iron ores		...	...	...	...		...	...	...	...		...	...	...	...
132	Non-ferrous metal ores,excl.uranium,thorium		76	76	76	76		...	...	...	...		...	...	...	...
141	Quarrying of stone, sand and clay		18	18	18	18		...	...	...	...		...	...	...	...
142	Mining and quarrying n.e.c.		12	12	12	12		...	...	...	...		...	...	...	...
C	Total mining and quarrying		107	107	107	107		...	...	...	...		...	...	...	...
401	Electricity production, collection, distribution		...	...	...	...		...	...	...	...		...	...	...	...
402	Gas production and distribution		1	4	1	1		...	...	...	...		...	...	...	...
403	Steam and hot water supply		...	...	...	...		...	...	...	...		...	...	...	...
410	Water collection, purification, distribution		6	4	6	6		...	...	...	...		...	...	...	...
E	Total electricity, gas and water supply		7	8	7	7		...	...	...	...		...	...	...	...

Zimbabwe

ISIC Revision 3		Output at basic prices					Value added at basic prices				
			(thousands of US Dollars)					(thousands of US Dollars)			
ISIC	Industry	Note	2010	2011	2012	2013	Note	2010	2011	2012	2013
101	Mining and agglomeration of hard coal		...	...	...	...		...	...	...	...
102	Mining and agglomeration of lignite		...	...	...	...		...	...	...	...
103	Extraction and agglomeration of peat		...	...	...	...		...	...	...	...
111	Extraction of crude oil and natural gas		...	...	...	...		...	...	...	...
112	Oil, gas service activities		...	...	...	...		...	...	...	...
120	Mining of uranium and thorium ores		...	...	...	...		...	...	...	...
131	Iron ores		...	...	...	...		...	...	...	...
132	Non-ferrous metal ores,excl.uranium,thorium		513222	190489	182336	265097		341586	112775	104561	156156
141	Quarrying of stone, sand and clay		19150	48658	21715	29451		12274	29582	12342	15176
142	Mining and quarrying n.e.c.		...	...	...	...		...	...	...	...
C	Total mining and quarrying		564642	271417	236321	326818		386130	174627	149172	203602
401	Electricity production, collection, distribution		...	...	...	...		...	...	...	...
402	Gas production and distribution		...	...	...	...		...	...	...	...
403	Steam and hot water supply		...	...	...	...		...	...	...	...
410	Water collection, purification, distribution		108832	275636	...	...		73006	198522	...	...
E	Total electricity, gas and water supply		108832	275636	...	...		73006	198522	...	...